Gender, Sex, and the Postnational Defense

OXFORD STUDIES IN GENDER AND INTERNATIONAL RELATIONS

Series editors
J. Ann Tickner, University of Southern California
Laura Sjoberg, University of Florida

GENDER, SEX, AND THE POSTNATIONAL DEFENSE

Militarism and Peacekeeping

ANNICA KRONSELL

OXFORD
UNIVERSITY PRESS

Oxford University Press, Inc., publishes works that further
Oxford University's objective of excellence
in research, scholarship, and education.

Oxford   New York
Auckland   Cape Town   Dar es Salaam   Hong Kong   Karachi
Kuala Lumpur   Madrid   Melbourne   Mexico City   Nairobi
New Delhi   Shanghai   Taipei   Toronto

With offices in
Argentina   Austria   Brazil   Chile   Czech Republic   France   Greece
Guatemala   Hungary   Italy   Japan   Poland   Portugal   Singapore
South Korea   Switzerland   Thailand   Turkey   Ukraine   Vietnam

Published by Oxford University Press, Inc.
198 Madison Avenue, New York, NY 10016

www.oup.com

Oxford is a registered trademark of Oxford University Press

Library of Congress Cataloging-in-Publication Data
Kronsell, Annica.
Gender, sex and the postnational defense : militarism and peacekeeping / Annica Kronsell.
   p.   cm.—(Oxford studies in gender and international relations)
Includes bibliographical references and index.
ISBN 978-0-19-984606-1 (hbk. : alk. paper) 1. Women and war.
2. Peacekeeping forces—Women. 3. Sweden—Armed Forces—Women.
4. Sweden—Armed Forces—Afghanistan.
5. Peace-building—Women. 6. Women and peace. 7. Militarism—Social aspects.
8. Women and the military—European Union countries.
I. Title. II. Title: Militarism and peacekeeping.
U21.75.K76   2012
355.0082—dc23   2011030905

1 3 5 7 9 8 6 4 2

Printed in the United States of America
on acid-free paper

CONTENTS

ACKNOWLEDGMENTS

This book is the result of a decade of research during which I have received generous support from very many people in many different ways and in different stages of the book project. Although I take sole responsibility for the content, it would never have been possible without your help. I want to thank all of you who have encourage and supported me, stimulated my thinking on the issues that are treated, given me constructive criticism, and helped in the completion of the text.

I want to recognize the importance of the Feminist Theory and Gender Studies section of the International Studies Association. I have benefited from the company of the sections' scholars in innumerable ways over the years. My special thanks go to Brooke Ackerly, Kathy Davies, Cynthia Enloe, Jan Jindy Pettman, Lisa Prügl, Laura Sjoberg, Christine Sylvester, Ann Tickner, Jacqui True, and Gillian Youngs. I have also enjoyed the wonderful encouragement of the Nordic network of feminist IR scholars and want to particularly thank Maud Eduards, Lene Hansen, Elina Penttinen, Maria Stern, and Tarja Väyrynen.

In particular, I want to acknowledge the University of Washington in Seattle for hosting me on various occasions. I am especially grateful to Professors Christine Ingebritsen and Terje Larsen at the Department of Scandinavian Studies who encouraged me to write this book and generously invited me to their department and to the community of Scandinavianists in the US. My thanks extend also to Nancy Hartsock and Elizabeth Kier at the Department of Political Science.

I am indebted to many scholars working in Sweden like Jeff Hearn and Lina Sturfelt and not the least to my own department where many have enthusiastically followed my research. I want to particularly mention Catarina Kinnvall and Karin Aggestam. Erika Svedberg at Malmö University College is my dear friend and colleague. We have discussed the issues of this book deeply and intensely over the years. Erika has taught me to go beyond the easily observable, to pay attention to the complexity of issues and study them deeply to find patterns, links, and power relations not otherwise noted. Thank you. I am also indebted to two anonymous reviewers for their valuable input and Angela Chnapko and Lora Friedenthal at OUP for their diligent work with the manuscript. I am grateful to my family who has been extremely patient and provided that private space of love, understanding, and comfort necessary for a productive research life.

LIST OF ABBREVIATIONS

ESDP	EU's Security and Defense Policy
EU	European Union
EUFOR	European Union Force
EUSEC	EU Advisory and Assistance Mission for Security Reform
EUMC	EU Military Committee
EUMS	EU Military Staff
IR	International Relations
ISAF	International Security Assistance Forces
LGBT	Lesbian, Gay, Bisexual, and Transgender
MONUC	United Nations Organization Mission in the Democratic Republic of the Congo
NAP	National Action Plan
NATO	North Atlantic Treaty Organization
PRT	Provincial Reconstruction Team
PSC	Political and Security Committee of the Council
PTSD	Posttraumatic Stress Disorder
SAF	Swedish Armed Forces
UN SCR 1325	United Nations Security Council Resolution 1325 on Women, Peace and Security (2000)

Gender, Sex, and the Postnational Defense

Introduction

In this book, I am interested in exploring the postnational defense and its gender implications. A postnational defense is one that pays less attention to the defense of the territory and more to the security situation outside its borders, often in cooperation with other states. Sweden is an example of a state that has moved from a national to a postnational defense, as is the European Union (EU), once a trade organization that has developed its own security and defense institutions. The postnational defense is the result of a longer process of changes in the security, defense, and military relations in Sweden, the EU, and the world but was articulated, for example, in the "Statement of Government Policy in the Parliamentary Debate on Foreign Affairs" in February 2010 when Swedish Foreign Minister Carl Bildt declared, "In a globalised world, foreign policy risks know no borders. Threats that originate far away from Sweden can be just as palpable as those that are rooted in our geographic proximity. . . . Our foreign policy is a part of the common European foreign policy" (9) and "membership in the European Union means that Sweden is part of a political alliance and takes its share of responsibility, in the spirit of solidarity, for Europe's security" (3). Hence, postnational security is envisioned as multinational and as achieved in solidarity with others well beyond the borders. Foreign Minister Bildt continued: "Our commitment in Afghanistan is based on our steadfast determination to help the Afghan people build up a functioning state" (9). The notion is that solidarity is not only with other states in Europe but with other people in faraway places. The development toward a postnational defense should be understood in the broader context of changed security relations.

Many scholars have argued that the end of the Cold War and the beginning of the war on terror have prompted dramatic changes in the global security context; war itself has been transformed and so have defense and military activities (Coker 2008; Ignatieff 2000; Kaldor 2001; Münkler 2005; Shaw 2005). These transformations in security, defense, and military relations signal a diminishing role for the nation-state in favor of multilateral security activities and a diversification of militaries' activities. Security is no longer associated only with states and nations; other understandings of security have become increasingly relevant. The notion of human security has gained broad recognition as part of the global security agenda. It is mirrored in the resolutions of the UN Security Council—for example, in resolutions on the responsibility to protect and on gender and peacekeeping—that recognize the importance of human rights and human security. An overarching question for this book is whether the broader recognition of human security, and in particular gender and security, has influenced the way the defense and military is organized and, if so, how this has occurred.

One reflection on the changed security context is that there is a divergence in military purpose. In turn, this is assumed to be a reaction to a transformation of war, away from state-controlled wars and military organizations to the dominance of internal conflicts—violent struggles that are often ethnified, have religious overtones, or are conducted over resources. Such changes are reflected in the nature of warfare and soldiering but also in constructions of nationhood and citizenship. Changes in the global security makeup have made governments as well as international institutions rethink defense strategies. For many of the liberal democratic countries in the rich north, it has implied a multilateral agenda. Strategic defense decisions, investments, and organizational changes have taken place among governments, and many have incorporated a peace agenda in their military organization. The development of two distinct types of state militaries exemplifies this: armed forces that engage in the 'war on terror' and those that engage in 'peacekeeping' efforts. The latter applies to Sweden and the EU, the cases developed in this study.

Afghanistan is a site where both types of militaries are present, side-by-side, in the same place. The U.S. Armed Forces are engaged in the war on terror and war-making activities against the Taliban forces. The International Security Assistance Forces (ISAF)[1] are there for peacekeeping purposes, to disarm and enable a secure environment fit for civilian activities. The ISAF was created through a UN mandate[2] and is led by the North Atlantic Treaty Organization (NATO). This uneasy relationship between peace and war-making is paradoxical for the defense organization but also for the people 'attacked' or 'protected' by these forces. All are militaries, but some are more prone to violence and regular war-making, such as the U.S. forces, while others are expected to come with security and peace, such as the Swedish forces. Afghanistan is a place that epitomizes contemporary global security. The contradiction of war-making/peacekeeping occurring at the same time and in the same place reflects this change in security and defense politics in the global context. Although militaries have diversified, it is the state military that continues to provide the major portion of the tools and organizations used to fulfill a security and defense purpose. This is the case in multilateral peacekeeping forces but also in EU's defense efforts. Militaries are organizations connected with the state, nation, and civic identity as well as militarism. This is crucial to understand the paradox that arises from the present predicament—that militaries that are trained for war now are to do peacekeeping.

Despite considerable transformation, much seems to remain the same. Continuity seems particularly evident regarding gender relations. The main question of this book is what role gender plays in all this; it asks: To what extent have gender relations been transformed in this postnational security and defense context? A focus on gender is highly interesting in a study on military and defense transformation because, on all levels of military, defense, and peacekeeping activities, ranging from the common soldier and conscript to officers, decision makers, and ministers, the gender gap is striking. In all these positions, men dominate and have dominated historically, in some instances completely. Traditionally, military, defense, and security activities have been associated with one gender (male) and with masculinity and heterosexuality. This is an historic phenomena also taken as a given. As such, gender analysis was perceived as uninteresting to military, defense, security, and international studies research until the early 1990s, when feminist international relations (IR), as well as some military scholars, began to contribute with interesting research to this field. In reflecting on the end of the Cold War, Cynthia Enloe (1993) shows that in processes of military, security, and defense transformation, gender relations also change and that in times of transformation, gender relations are often challenged and reconstructed (Enloe 2000). This book aims to make a contribution to feminist IR and feminist security studies by asking: In which ways are gender relations changing in relation to the emergence of the postnational defense?

United Nation Security Council Resolution 1325 (UN SCR 1325) on gender and peacekeeping from 2000 is crucial in this new security context and points to an element of change regarding gender and security. The international order that emerged after the fall of the Berlin wall in

1989 opened up the possibility to rethink the meaning of security. As human security became a feature of the Security Council and the larger UN, gender became relevant. A network of women's groups and feminist activists working around the UN, who had expertise in this area, were able to influence the Security Council (Kronsell 2010). As we will see in this book, UN SCR 1325 has been crucial to the development of gender awareness and policies in the practice of the Swedish defense organization. Also, when the EU or individual member states call for gender mainstreaming in peacekeeping, they usually refer to this resolution. In this sense, this book ties into and expands on the research on the UN SCR 1325 by Laura Shepherd (2008) and many others (e.g., Cohn et al. 2004). UN SCR 1325 together with other policies and equality norms make it increasingly difficult for militaries to continue politics as usual. Such gender mainstreaming strategies put the gender effects of military activities in the spotlight.

Much has been written on the United States and the militaries engaged in the war on terror; some work has focused on gender and feminism. Riley et al.'s (2008) book *Feminism and War: Confronting U.S. Imperialism;* Laura Sjoberg's (2006) *Gender, Justice and the Wars in Iraq;* and Cynthia Enloe's (2010) *Nimo's War, Emma's War: Making Feminist Sense of the Iraq War* are three fairly recent examples. Over the years many feminist researchers have contributed to the understanding of military institutions—how they reproduce gendered militarism and how women are treated; examples are Stiehm (1982, 1989), D'Amico and Weinstein (1999), and Carreiras and Kümmel (2008a). Peacekeeping, as the activity that many militaries are involved with today, has also been explored in the literature. Louise Olsson and Torunn Tryggestad's (2001) edited volume *Women and International Peacekeeping* focuses particularly on the role of women in international peacekeeping; Mazurana et al.'s (2005) *Gender, Conflict, and Peacekeeping* investigates the meaning of gender in peacekeeping; and Sandra Whitworth's (2004) *Men, Militarism & UN Peacekeeping* is a feminist analysis of peacekeeping, paying particular attention to the way masculinity and militarism are embedded in peacekeeping; Canada is the case study. This book contributes to those research fields by studying the EU and by examining in-depth research on Sweden's defense transformation. The EU, of which Sweden has been a member since 1995, has agreed on a common security and defense agenda that has a strong emphasis on peace enforcement, peacekeeping, and crises management. Although some of the member states, particularly the United Kingdom but earlier also Italy and Spain, have taken a more active part in the war on terror, Sweden has not. Sweden is militarily nonaligned and not a NATO member.

Sweden's military and defense forces have responded to the new security context. What is most notable is the turn away from neutrality as a security doctrine and the abandonment of the territorial defense. To understand what the transformation toward a postnational defense implies for defense and military activities requires a deeper case study. An in-depth analysis of the transformation of the national defense is, in this book, carried out through the study of the Swedish Armed Forces (SAF). In many ways, it is a symptomatic case for the gender-aware postnational defense. It provides important general knowledge about gender dynamics in relation to nation-building and citizenship. It discusses how Sweden's defense forces have become geared toward peacekeeping through their engagement with the ISAF in Afghanistan. The European defense transformations are considered as well. In the EU, there has been a united undertaking promoting a common security and defense agenda. The tendency is similar here, too, for peace and war tasks collapse into one: EU Battlegroups use military means that become the tools to carry out peace tasks. The EU Battlegroup is a military unit expected to enforce peace through military means. Through these specific cases, this study probes the question of what changes in gender relations come with defense and military transformation.

THEORETICAL FRAMEWORK

What analytical tools are available for the study of gender and sex in postnational militaries and peacekeeping? The theories that inform the analysis of this book endorse a constructivist perspective. Feminist theorizing is concerned with the social construction around sex difference and considers gender to be a main organizing principle of social relations. This implies that a focus on gender could be relevant for every social relation and organization. Gender understood as socially constructed is, as Ann Tickner (2006, 24) suggests, a "variable category of analysis to investigate power dynamics and gender hierarchies." In this study, feminist theory informs the critical analytical position that serves as a base to extend the scope of theory but also to critically challenge the concepts on which theories and phenomena are based. How gender can and should be understood and studied is the basic question for feminist research. Traits and behavior historically associated with men and women as bodies often characterize the public debates on gender. Such debates are often highly simplistic. A reoccurring example is the common tendency to equate gender with women. Feminist theorists insist that gender is highly complex. One of the ambitions of this book, which is also a contribution to feminist theory in general terms, is to show the complexity of gender in the context of military and defense relations. That gender is complex also means that it is understood in different and contested ways among feminists.

Starting with a constructivist perspective on gender, this book's central focus is on gender constructions in the military and defense field as they become evident through analyses of texts, narratives, and images. Gender constructions are only reflections of what is done, lived, and performed in daily practice. In this sense, gender is intimately connected to the materiality of daily life, in doing things such as cooking, driving tanks, loving, protecting, or even killing. Joan Acker (1992, 250) writes: "Gender refers to patterned, socially produced distinctions between female and male, feminine and masculine. Gender is not something that people are . . . it is a daily accomplishment that occurs in the course of participation in work organizations as well as in many other locations and relations." Judith Butler (1990, 25) approaches gender as performativity—produced through performance—or by the 'doing': Gender is "constituting the identity it is purported to be" and "there is no gender identity behind the expressions of gender." This means that gender is a continuously (re)constructed category and that gender construction takes place in the continuous processes and activities of daily life. Gender subjectivities, or masculinities and femininities lived by men and women, are constructed in relation to one another and to other social categories. Thus, in the doing of gender, the differentiation between masculinity and femininity is a recurring element. Masculinity has no meaning without femininity. They depend on one another for their definition, yet they are not of equal worth and status. It is also in this relationship that gender power is constituted and, thus, gender analysis often pays attention to interactions between individuals. In the defense and military context, sex and sexuality play an important role in the construction of masculinities and femininities. Security, defense, and military institutions have carried out their activities based on a norm of heterosexual masculinity.

The researcher cannot study 'the doing of gender' per se, because it is subjectively lived and done, but can only get snapshots of gender constructions, for example, via texts or images. A limitation is that it tends to project gender as static, which it is not. Through 'doing gender', individuals form subjectivities in which they internalize, reify, or resist gender norms. The 'doing' takes place in a context of gender norms. This gender order is a power order because the relations are asymmetric. Different power relations intersect and overlap, such as those relating to class, sexuality, race, ethnicity, and age, hence adding to the complexity (Lykke 2010; Peterson 2008; Walby 2009). This complexity implies that there is not one power order but intersections of power, and it suggests that the intersections of power can be studied in various ways (Ackerley et al. 2006). For pragmatic reasons and to achieve analytical stringency, gender is the main theme here.

The power asymmetries associated with the gender order have often been the focus of feminist critique and concern. The gender order is expressed in terms of gendered material divisions, such as the representation of men and women, or wage differences. This book takes material aspects into account. The gender order is also expressed in norms and in a symbolic sense, as images of gender, for example, with the meaning of 'mother' to the nation. In the context of the defense, the analysis is directed at what is perceived as normal and normative: male bodies, masculinity, and heterosexuality. In this study we pay particular attention to the images and conceptualizations of gender and sex that are taking shape in the post-national security and defense context. What type of gender images are produced as a result of the work with human security?

This book takes as a starting point a constructivist view of gender relations, but with it comes also an institutional perspective. Based on the notion that institutions matter, this approach argues that organizational rules, norms, and features influence actors, and this has political outcomes. Furthermore, a constructivist perspective on institutions considers norms an integral part of the construction of subjectivity, interest, and meaning in political life (cf. Risse 2004, 163), a view endorsed by feminist constructivist IR scholars (Locher 2007; Locher and Prügl 2001). Hence institutions, such as militaries, are not merely 'intervening variables' but a part of the construction of a gender order and gender subjectivities. The arguments coming out of various institutional perspectives are that actors will moderate their strategies and behavior to fit with institutional features (cf. March and Olsen 1989; Peters 2005). Institutions, on their part, tend to have particular, 'pattern-bound' effects over time, caused by locking into place and creating certain rules and decisions and thereby contributing to path dependencies. It provides an explanation as to why historic or traditional notions of gender seem so resilient. Institutions are thick, and norms are embedded. The dominance of heterosexual masculinity as a norm is an example of an embedded norm system, highly institutionalized over a long historic period. In institutions that have historically been governed by men, the male-as-norm dominates, and gender and sexuality are quite clearly important institutional resources. Militaries are examples of institutions in which core practices (such as military cohesion) are constituted on gendered norms associated with masculinity and heterosexuality. These are the kind of institutions surveyed in this study. Although institutions can be understood more broadly, the focus here is on political and operational institutions in the field of security and defense and, hence, closely connected to the state.

Constructivist approaches in IR have commonalities with feminist agendas because, as Locher and Prügl (2011, 114) argue, they share an ontology of becoming—seeing the world not as one but as one becoming. Institutions can both enable and constrain, and they not only delimit our world but also systematically distribute privilege and thereby create patterns of subordination. From a feminist constructivist perspective, institutions and organizations are seen as deeply embedded in their environment. "Embedded" implies that the norms, values, and procedures expressed in an institution, formally or informally, form a part of a larger spectrum of power relations predominant in the time and space in which the organization is situated. Social orders such as gender orders are often encompassing, have distributive effects, and privilege certain groups over others. Power works both discursively and in institutional practices. The discursive is expressed in communication, symbols, images, and texts, while institutional practices are expressed in procedures and behavior. In this study, my aim is to capture the constructions around gender and sexuality in the postnational defense through this constructivist institutional approach.

Through institutional practices and processes, norms are enacted and reproduced. Particularly interesting for gender is the way institutions relate to gender subjectivity. Institutions make certain subjectivities possible (and not possible). March and Olsen (1989, 39–52) call this "the creation of meaning" and explain that individuals who engage in institutional activities tend to search for meaning and a sense of purpose in what they do in their daily work. In going about their daily activities, individuals appeal to institutional norms and also do or perform gender practices, reproducing the male-as-norm and/or challenging it. This is the dual character of institutions. Gender subjectivities are constructed (but also resisted) through the performance of institutional tasks and procedures. As we see evidence of in this book, to form a feminine subjectivity, for example as a 'women-in-arms', is problematic in an institutional context in which the dominance of masculinity makes such a subjectivity a very uncomfortable fit with military practice. Thus, to capture gender constructions, it is crucial to study institutional practices where gender and sexuality are performed.

In understanding gender relations, feminist theories are particularly relevant. Feminist research is interdisciplinary and spans many research fields. A broad selection of feminist work informs this book, mainly elaborated in the areas of international relations and organizational and institutional theories. Work in the field of critical masculinity studies is also highly relevant (Barrett 2001; Connell 1995; Gardiner 2001; Hearn 2004; Hearn et al. 1989; Kimmel et al. 2005; Whitehead 2002) as it investigates gender issues with a specific focus on men and masculinity. Critical masculinity studies employ feminist theory, and the two fields have a common objective: to deconstruct gender relations with a vision to change gender stereotypes. This is developed in greater depth when theoretical insights from masculinity studies are used to discuss empirical cases and examples in the various chapters. The critique of hetero-normativity in the queer theory literature and its analysis of heterosexuality are also applicable, but here is limited to only to a few scholars (e.g., Butler 1990). This is because queer theory tends to focus on literary and media critique and less on institutional aspects, militarism, war, or international relations. Specific studies that deal with the role of LGBT persons in the military are included.

The concept of militarism is also important. Military organizations are part and parcel of a norm system that sees war-making as a legitimate way to resolve problems and conflicts (Fogarty 2000; Jabri 1996; Skjelsbaek 1979). This norm system is often called *militarism*. A focus on militarism suggests that there is a deeper conceptual construction involved in militaries, defense politics, and war-making. Militarism provides a different perspective than what IR realist analysts, who dominated defense politics during the Cold War, argued about war. War was seen as the outcome of rational decisions based on state's interests and in Clausewitzian terms as politics by other means. While the military machinery and the power to inflict violence is the hard power of a defense politics, militarism constitutes a form of soft power, in Joseph Nye's (2004) terminology. Michael Mann's (1987, 35) definition of militarism—"as a set of attitudes and social practices which regards war and the preparation for war as a normal and desirable social activity"—is quite useful here. Enloe (2007, 4) is more concrete when she describes militarism as the belief that hierarchy, obedience, and the use of force are particularly effective in a dangerous world.

Different versions of militarism exist over time, across societies, and via different types of organized violence, but the common element is this taken-for-grantedness of war in politics. The concept of militarism is commonly perceived in pejorative terms, as it is often used to denote excess. This is exemplified in Johnson's (2004, 23) study of U.S. militarism in relation to the politics of empire and in Shaw (2005, 95) who detects militarism when war preparations flow over into society. Military historian and strategist Martin van Creveld (2008, xii) endorses this view and is disturbed by the fact that academics who look at what he calls "the culture of war . . . as an expression of that worst of all bad things, 'militarism'." While he glorifies war by saying that "War, and combat in particular, is one of the most exciting, most stimulating activities that we humans can engage in" (van Creveld 2008, 411), he also recognizes that feminism is one of its most serious threats in two ways. Allowing women in the military and in combat will destroy militarism, and women may turn against this culture and thereby ridicule it (van Creveld 2008, 409).

Indeed, women and feminists, particularly those who have an origin in the women's peace movement and call themselves antimilitarists, are very critical of militarism. Exemplifying such a highly critical view is Ann Scales' (1989, 26) definition of militarism as "the pervasive cluster of forces that keeps history insane: hierarchy, conformity, waste, false glory, force as the resolution of all issues, death as the meaning of life and a claim to the necessity of all of that." The reason many feminists are highly critical of militarism is due to the relationship between the gender order and organized violence. Betty Reardon (1996, 5) connects militarism with the gender order; she uses the terms *sex system* and *war system* and says that they are interdependent because both manifest social violence. Military values support the use of coercive force in the interest of the nation, while the gender order is based on the use of or the threat of violence. Furthermore, militarism manifests extreme or excessive forms of masculinity, argues Reardon (1996,15). Based on evidence from Northern Ireland, Israel, and former Yugoslavia, Kaufman and Williams (2007, 203) claim that there is a correlation between militarism and violence against women. Kimberly Hutchings (2008, 389) explores the connection between war and masculinity and contends that "qualities such as aggression, rationality, or physical courage are identified both as an essential component of war and also of masculinity at a given place or time"; thus, there is a functional relationship between masculinity and militarism. Cynthia Cockburn (2004, 44) develops this further, through the notion of a 'continuum of violence' whereby she argues that both systems—the gender order and militarism—rely on structural violence, threat and domination, discipline, and hierarchy: "The power imbalance of gender relations in most (if not all) societies generates cultures of masculinity prone to violence. These gender relations are like a linking thread, a kind of fuse, along which violence runs. They run through every field (home, city, nation-state, international relations) and every moment (protest, law enforcement, militarization), adding to the explosive charge of violence in them." Hence, there is an important connection between military violence and gender violence. McCarry (2007) also talks about two interrelated functions of male violence. On an individual level, men use violence to control women and children. On a structural level, male violence "has the effect of perpetuating a system of male domination" (McCarry 2007, 405). The international system is, thus, a structure of male domination in terms of states led by male elites (cf. Wadley

2010). In that structure, war is a form of interaction and militarism the norms that can justify that interaction.

The feminist take on militarism proposed in this book is to look at gender relations from within the defense institutions, for example, by examining how masculinities are constructed in relation to military and peacekeeping practice or what happens when women claim to take part and gender issues are imposed on what would be a masculine militarist institution. Gender and sexuality as well as other differences are used as resources in militarism and military practice in various ways. On the relationship between masculinity and militarism, Higate and Hopton (2005, 436) say that militaries provide means by which values and beliefs that are associated with masculinity are both "eroticized and institutionalized" and have "the potential to link masculinity with the political concerns of the state." So we can say that masculinity is an integral part of militarism or military culture.

For these reasons, feminist scholars have been highly critical of militarism and militaries; many are antimilitarist, pacifists, and against violence in all forms and see this stance as a central element of being a feminist activist. There is also another line of research within feminist IR theory that remains critical to the gendered aspects of militarism, defense, and military organizations but nevertheless does not see feminism as necessarily pacifist. Laura Sjoberg (2006) is such a scholar who argues that it is necessary to recognize that situations so cruel and horrible may occur (e.g., genocide) and that remaining passive would be more unjust than using military means to try to advert a disaster. Thus, she argues, there is a need for an ethics of war that is feminist.

Scholars in the tradition of just war theory have argued precisely that wars are sometimes necessary, but when they are, morals and principles must be applied as to not let everything go to brute violence and massacre (Aggestam 2004; Elshtain 1992; Waltzer 1977). The need for an ethics of war that can help us act courageously is all the more necessary in a time when risk and fear dominate the international security agenda, says Christopher Coker (2008). The concern with the ethics of war in relation to militaries has been taken up also by Mary Kaldor (2001). She calls for a cosmopolitan approach and suggests that cosmopolitan values are important in meeting the challenges of the 'new' wars. In this, cosmopolitan troops would be the "legitimate bearers of arms" (Kaldor 2001, 131). The literature on cosmopolitan militaries is interesting to this study (Elliott 2002; Elliott and Cheeseman 2004). First, such militaries are postnational because they are geared toward 'saving distant others' in the name of human rights rather than serving the nation and the national territory.

In practice, cosmopolitan militaries can be expected to be actively involved in peacekeeping activities. Second, the military organization is simultaneously transformed internally as the organization is increasingly subject to democratic norms and the rule of law in order to reflect civilian norms and be able to set an example of a new cosmopolitan organization where they are deployed. Two ideas follow from this. First, equality policies, gender mainstreaming, and antiharassment policies can be expected within cosmopolitan military organizations, and the division of labor and distinction between civilian and military tasks become less relevant. Second, from this follows other expectations on individual soldiers as they take part in cosmopolitan military activities. The theory on cosmopolitan militaries articulated this way is interesting to feminist scholars because the changes expected from this transformation point in the direction of a denationalization—with the emphasis on saving distant others—and a demilitarization—when civilian, democratic values and norms become more important. This seems highly relevant for a feminist agenda.

This section has provided a loose theoretical framework or a set of theories that are used in the analysis of the cases in this book. The goal is to apply these theories to generate empirically derived insights to the fields of literature in gender studies and thereby learn more about how gender and sexuality are reflected in defense and nation-building, in the military organization, in peacekeeping as a new defense and policy strategy, in peacekeeping practice, and in the making of a new defense institution.

RESEARCH MATERIAL

The material that provides the empirical base for this study was gathered from 1999 to 2011.[3] The kind of empirical material used varies from official governmental reports, parliamentary debates, motions and committee reports, official documents such as annual reports, and equality plans from governmental institutions, such as the Armed Forces, Service Administration, and Crises Management Agency, to EU and UN documentation, official documents related to the development of the European Defense and Security policy, and reports from member states. Studies and reports issued by the Swedish Defense College, the Swedish Defense Research Agency, and the Institute of Security Studies have also been important, as is information material directed to the public, including web pages, information folders, debate articles, newspaper articles, documentaries, and recruitment ads. The material also consists of various internal newsletters and reports from the organizations, often directed to organizational members or staff in the wider area of defense activities. This material has been complemented with more than 30 interviews, on-site observations, and field studies. A major part of the material is used as background sources with specific references only when directly applied in the text. The research on Sweden and the EU is compared with scholarly studies, reports, and data from other similar cases among European states and Canada.

OUTLINE OF THE BOOK

The national military organizations in Sweden, as with the rest of Europe and most of the world, are historically well-established state institutions with strong connections to nation-building processes (Tilly 1990) and to the heterosexual male citizen soldier (Goldstein 2001). The trend since the end of the Cold War is that the national defense of the EU member states has become postnational and the model of the soldier based on the heterosexual male soldier has been challenged. The call for changes was brought on by the transition to a post–Cold War era but also by the demands for democratization, equal treatment in civil society, and political life. While it is argued here that this is the trend for Europe there is a parallel trend in and increased emphasis on national identity in other parts of the world—struggles often associated with militarization. An understanding of processes of nation-building in relation to military and defense politics remains highly relevant as a background to understand the postnational defense and so too is the gendered analysis of nation-building and its relation to military organizations, security, and defense. It is suggested here that there is something important to learn from a feminist analysis by looking at the case of Sweden and the EU.

Chapter 1 shows how Sweden's defense organization has been implicated in a type of gender politics that has constructed a specific gendered national identity as well as a gendered citizenship. *Folkhemmet* is conceptualized as a uniting theme for nation-building. The home and the nation were joined through the notions of 'mother' and 'motherhood', while the "folk" where men and women with highly differentiated tasks and duties. National identity was connected to the state through specific institutionalized relations and practices, and in this sense corresponds to what Yuval-Davis (1997, 1) has said more generally about constructions of nationhood: that they normally involve notions of man- and womanhood. The empirical focus of this chapter is on conscription and the voluntary defense organizations and their part in the process of gendering the nation. Through analyses of these institutions, it is shown that embedded in their practices are clearly differentiated expectations of male and female contributions to the state and the nation. The defense sector is deeply implicated in this. Often, highly stylized constructions of men and women surfaced. Albeit symbolic, they do have substantial effects on women's and men's status as citizens and on their role as political subjects. Conscription and voluntary defense organizations are analyzed to show in some detail how gender notions of citizenship were inscribed in practice. While gender imaginaries of 'just warriors' and 'beautiful souls' are articulated on a symbolic level, the analysis shows how gender constructions are expressed in institutional practices, such as those relating to the information function, how security is perceived, and what is expected of members in voluntary defense organizations. Thus, chapter 1 focuses on the backdrop of the postnational defense. In addition, an analysis of the gender of nation-building through the defense and military is necessary to understand the transformation to a postnational defense and military. The transformation taking place is conditioned by this.

Chapter 2 uses institutional perspectives to analyze norms of masculinity embedded in the military institutions and explores gender and sexuality in three ways. First, it discusses how norms of militarism and masculinity shape and constrain the possibility for women to fully become soldiers and attain a women-in-arms identity. Second, the analysis looks at gender as an organizational resource. It shows how sexuality plays a significant role in institutions dominated by masculinity. It is about the construction of heterosexual masculinity in relation to homosociality, argued to be highly relevant for group cohesion, a key feature of military performance and an explanation as to why military policies exclude women from combat and homosexuals from serving in the military. The chapter ends with an assessment of strategies promoted by the Swedish Armed Forces (SAF) in recent years. The strategies aim to deal with intolerance, harassment, and discrimination within the military organization with the aim to increase gender equity and eliminate discrimination and intolerance. This is also a promotion of ideas that contrast the norms embedded in the military institutions, which nevertheless fit with a cosmopolitan view of military practices, elaborated in the following chapter. From this reading it is possible to see how gender and sexuality—or those 'differences'—play out against deeply embedded norms of military heterosexual masculinity.

Chapter 3 reflects on Sweden's development toward a postnational defense and asks whether the changes observed in the military can be understood through the notion of cosmopolitanism. It is compelling to a feminist analysis because cosmopolitan ideals propose to democratize and demilitarize contemporary militaries (Elliot and Cheeseman 2004).

The Swedish postnational military indeed emphasizes peacekeeping activities with a vision to save distant others in the name of human rights. The Swedish military appears to be pushing a gender-friendly image toward the world. It is a way to live up to cosmopolitan ideals and to gain legitimacy and popularity domestically while at the same time confirming Sweden's image in the world as an internationalist, progressive, gender-friendly state. While the case study is Sweden, the chapter shows that the findings have relevance also for other peacekeeping nations. Simultaneously, as Sweden engages in peacekeeping abroad, it carries out reforms and introduces democratic and civilian values. This may drastically change the military organization as well as the expectations on the soldier/peacekeeper. The chapter asks: What are the implications of this? To turn an organization that historically has dealt only with war into an organization that has making peace its goal is not easy. As it turns out, this is associated with another function of militaries: to make boys into men. The emphasis on peacekeeping tasks certainly challenges the military's ability to perform this function. Finally, the chapter suggests that feminist notions of emphatic cooperation and ethics of care can be a way to complement cosmopolitan values and guide peacekeeping activities.

Chapters 4, and 5 develop the notion of the cosmopolitan, postnational military by looking specifically at peacekeeping operations in chapter 4 and the EU's defense and security policy and battlegroups in chapter 5.

Chapter 4 turns toward the internal practices of the military institution in a postnational defense that has as a main activity 'to save distant others in the name of human rights'—in other words, through peacekeeping and peace-enforcement tasks. The chapter shows how gender constructions and sexuality are dealt with both as problems and resources in the postnational military. It begins by discussing the ambition to rethink masculinity within the Swedish peacekeeping forces. It goes on to look at sexual misconduct in UN peacekeeping operations to illustrate how sexuality is perceived, both within the organization and in relation to constructions of masculinity. Sexual misconduct is understood as an individual problem, while the chapter argues that sexual misconduct is intimately connected to organizational practice and the context in which peacekeepers are stationed. The chapter studies constructions of masculinity mainly through its connection with sexuality and considers constructions of femininity in relation to peacekeeping activities. It does so by asking what women peacekeepers are expected to contribute. First, women are seen as the antidote to the problem of sexual misconduct, as the female presence gives (renewed) legitimacy to the mission. Second, female peacekeepers are seen as a resource because they are able to talk to and engage with local women, make possible the implementation of UN SCR 1325, and assure a more complete security situation. Here, female peacekeepers are valuable as those who gather information from women. This intelligence function is especially important in highly patriarchal settings. The chapter looks more closely at and builds on the example of the Swedish ISAF stationed in Mazar-i-Sharif in northern Afghanistan.

Chapter 5 takes its starting point in the development of the European Security and Defense Policy (ESDP). It may be expected that because the EU is a fairly recent institution and a civilian power, common EU defense arrangements would definitely represent a more civilian and postnational type of military, perhaps different from anything we have seen before. Some evidence of this trend is that EU missions are multinational and strictly about peace enforcement and peacekeeping; an example is the rapid deployment forces known as the battlegroups. The chapter asks: To what extent has and can this 'new' defense organization of the EU reformulate goals of militarism to become less national/more cosmopolitan and destabilize the male-as-norm for military activities? Is there any evidence that the ESDP is the result of a reflective process that has risen out of the ashes of the Cold War militaries of Europe? How have gender balancing and gender mainstreaming policies, already in place at the time the ESDP was created, been implemented and integrated into the institutions of the EU's security and defense?

The final chapter concludes the book by revisiting the main research questions in view of the findings in the previous chapters. It discusses the major findings in terms of what contributions they make to feminist theory in international relations, militarism and gender mainstreaming, and the study of masculinities.

CHAPTER 1

Mothers, Soldiers, and Nation in the 'Neutral' Defense

Inger Segelström is an influential social democratic politician. In an interview in 2001, I asked her whether she thought that military conscription as a requirement for men only was compatible with democracy and equal rights. She said: "Women do their duty when they carry the future generation for nine months. Pregnancy is women's duty toward society. It is a comparable sacrifice that women do for the nation. Why should women, if also conscripted into the military, have a double burden? That would be unfair" (my translation). This view represents a common perception on the way women's and men's duties toward the nation differ. It is typical in that it articulates the gender dichotomy associated with nation-making. This gender dichotomy appears strikingly similar across nations and over time. The mother and the conscripted solider have two distinct but complementary roles.

Nation is a process by which community is imagined and reimagined in the politics of nation-building (B. Anderson 1983). Collective identity formation takes place on a symbolic level; that is, it provides imaginaries about who the people are and what their roles as citizens are. Although imaginary and symbolic, such processes have material effects on state institutions' practices and citizens. The national defense is an important background against which to understand the challenges facing the contemporary postnational defense. Swedish military and defense practice was founded on gendered ideas about the processes of collective identity formation and nation-making and notions of citizenship. The aim of this chapter is to probe this position in some depth. The connection between the defense and the nation is illustrated by exploring the practices of military conscription and voluntary defense organizations.

During the Cold War, Sweden's national self-image consisted of two interconnected dimensions: *folkhemmet* and Sweden's projection of itself into the world (cf. Wæver 1999), expressed as the neutrality security doctrine. The concept of *folkhemmet* constituted a uniting theme for nation-building. It conveyed the societal model of the welfare state as a common home for all people, regardless of social position (e.g., Hirdman 1989; Lagergren 1999). Uniting people around the welfare state provided a common destiny important to Swedish nation-building (cf. Yuval-Davis 1997, 18–19). As Munn (2008, 443) explains, the objective of national identity-making is to accomplish the state, and it relies on a belief in a collective commonality. Thus, the Swedish national identity during the Cold War rested on the vision of *folkhemmet* connected to the state through specific institutionalized relations and practices, such as conscription and voluntary defense organizations. All processes of common identity formation require a certain ethnocentrism through which the nation can differentiate itself in the world. The second dimension of Sweden's self-image as a nation, Sweden's neutral security policy, particularly illustrates this and is discussed in chapter 3. It is important to understand the nation-building process and its gendered dimensions, as the postnational military and defense can be perceived as a break with this particular construction of nation—an issue probed in much of its complexity throughout this book.

WOMEN AS MOTHERS AND MEN AS THE PROTECTORS OF THE NATION

The main thesis of Yuval-Davis's (1997, 1) seminal work *Gender and Nation* "is that constructions of nationhood usually involve specific notions of both 'manhood' and 'womanhood'." Many scholars have discussed this theme and called attention to the importance of motherhood to nation-building. Dubravka Zarkov (2007, 20) says about the maternal body that it "is the body vested with the power to give birth to the nation. As such, it is both vulnerable and powerful, a potential target of attacks and a focus of protection, a fierce defender of its honor and its offspring." The association between nation-making and motherhood may be particularly accentuated and powerful in times of violent conflict but has general relevance also in peacetime. When observing military practices, defense, and war across time, a gender dichotomy prevails. It is, as Judith Hicks Stiehm proposed in 1982, a dichotomy related to one of the state's defining functions: the legitimate use of violence. The state's monopoly on the use of violence relies on two categories: the protector (male soldiers) and those to be protected (women and children). This dichotomy, she argues, has set the stage for a concrete state practice of forbidding or making it very difficult for women to act as the protectors or defenders, take to guns, or learn how to use them (Stiehm 1982). More recently, Iris Marion Young (2007) argued for the continued relevance of this dichotomy, which is highly visible in the context of the war on terror and the resulting heightened security concerns (Young 2007, 117–139). Jean Elshtain's (1965) work is useful here. She starts with the concept of 'just warriors' and 'beautiful souls' to describe the gender dichotomy that historically has dominated militaries and defense organizations. It is based on "a rough and ready division of male life takers and women life givers" (Elshtain 1995, 165). These are stylized ideas about what is expected of a man and a woman, respectively, in terms of their contribution to the state and the nation. Paraphrasing Elshtain (1995), the 'beautiful souls' are too good for the military but yet absolutely necessary to it. They (i.e., women) are situated in this scenario as a sort of civil cheerleader, inspiring the soldiers to fight well for a just cause, but they are always outside or on the margins of the military. Miriam Cooke (1996, 15–16) encapsulates this:

> The War Story reinforces mythic wartime roles. It revives outworn essentialist clichés of men's aggressivity and women's pacifism. It divides the world between the *politikon*, where men play 'political' roles, and the *oikon*, where women are lovers or mothers. The War Story proclaims that this sex segregation is justified for biological reasons: the men are strong; therefore they must protect the women who are weak. It is written in their genes that men shall be active and women passive.

Gender constructions are always relational. In the conceptual pair, referred to here as 'just warriors' and 'beautiful souls', one always needs the other to make the picture complete.

Because it is relational, the dichotomy implies that the characteristics, tasks, and behaviors associated with the pair are complementary. While complementary, the soldier and the mother share an interesting similarity in their relationship toward the nation, as pointed out by Elshtain (1995, 222–225). Nation is a powerful collective identity—so powerful that it has been able to call for significant sacrifice from its citizens, such as surrendering individual autonomy for the needs of the nation or even dying for it (e.g., Ritter 2002, 212). Both soldiers and mothers make sacrifices and go through liminal experiences, Elshtain (1995, 222) argues that "once you have given birth or been in battle, you are no longer the same person." Both soldiers and mothers are concerned, anguished, and traumatized by whether they have done enough, for the nation as soldiers or for the children/family as mothers (see also Sylvester 1994, 329).

Indeed, the idea of mother and motherhood has been of crucial importance in nation-building and key in the gender constructions associated with it (Albanese 2006; Kaufman and Williams 2007; Zarkov 2007). Maud Eduards (2007, 38–42) argues for the relevance of this to Swedish nation-building. The home and the nation are joined through notions of the mother and motherhood. This was expressed in relation to conscription practice in the argument of Segelström cited at the beginning of this chapter. This relationship is also evident in the symbolic expression of the Swedish nation associated with the maternal body and captured by the notion of 'Mother Svea'.

Historian Charlotte Tornbjer (2002) studied the construction of motherhood in relation to Swedish nation-building during the two world wars. She found that women, in the collective identity of the nation, were first and foremost constructed as mothers or mothers-to-be. However, how motherhood was conceptualized varied across time. To exemplify, Tornbjer discusses a range of mother images, including heroic, religious, victimized, peaceful, rational, and patriarchal motherhood. She concludes that although the expectations regarding the contribution of a mother to the nation vary, as is conceptualized through such different images of

Figure 1.1: Moder Svea. Statue on Stora Nygatan, Stockholm, from Wikipedia Commons.

motherhood, what remains constant is the tendency to see women as mothers in relation to the nation.

The nation is a symbolic expression but also becomes an order in which every person has a specific role and function to fulfill (Eduards 2007, 34) and thus is formed at the level of the individual, in the self, and makes possible certain subjectivities. The notion of citizenship is important here, and it regards the relationship between the individual and the state. Military conscription, for example, is a specific relationship between the male individual and the state. It defines male citizenship and a man's place in the nation. Because gender is relational, it simultaneously constructs women as 'other' and as different.

GENDERING *FOLKHEMMET* THROUGH MILITARY CONSCRIPTION

The strong identification of the military with the nation gives the armed forces a kind of influence and privilege not matched by any other societal institution (Enloe 2000, 46). Military institutions wield both hard power—weapons, equipment, and the possibility to inflict harm and even kill—and soft power, related to norms of militarism and nation. The following examples focus on military conscription because conscription into mass armies has been central for national military and defense organizations (Mjøset and Van Holde 2002, xiii; Tilly 1990). Karsten Friis (1999, 5) argues similarly about Norwegian conscription, saying that it is an institutional practice whereby the national meets the state. Insook Kwon (2000, 28) talks about conscription in South Korea and says that it "has played a crucial role in constructing citizenship, nationhood, masculinity" and has created "the essential glue" that bind these to the concept of nation. Bjørn Møller (2002, 279), speaking more generally about the historic connection between conscription and nation-making, says: "Conscription was a direct reflection of the new basis for state-building, namely nationalism, as the conscripted armies served as 'melting pots', bringing diverse strands of the population together and thus furthering a sense of community." Yet, only half the population was included in these nation-building efforts. For Sweden, conscription was also tied to the physical dimension of the nation: The main task was to maintain a defense organization that could be mobilized for the entire territory's protection.

In Swedish conscription, collective identities and individual identities were co-constituted and mediated by an individual sense of duty called 'the duty to defend' or *värnplikt*, as Erika Svedberg and I have shown in previous work (Kronsell and Svedberg 2006, 2001). A more or less formalized sense of duty of the individual vis-à-vis the state has been a central organizing principle of the Swedish welfare state, or *folkhemmet* (Trägårdh 1997, 253, 266). The traditional sense of duty previously associated with the family and the rural context was captured by the social movements of the early 1900s (Ambjörnsson 1998), became a matter of solidarity in the workers' larger collective, and provided the political incentive to realize the national project of *folkhemmet*. From this, a social contract between the state and the individual emerged (Olsson 1990). It was a gender-differentiated contract that provided the male citizen with certain duties and rights (cf. Pateman 1988). In conscription practice and in the voluntary defense service, the sense of community was achieved through blending of social groups around the defense 'duties' and 'contracts' and through education about security and defense.

Clearly, core ideas about gender that have defined the male and female citizens of Sweden are tied to the practice of conscription. The 'just warrior' "takes up arms reluctantly and only if he must to prevent a greater wrong or protect the innocent from certain harm": It is a tragic but necessary task (Elshtain 1995, 127). The conscript soldier can be associated with the 'just warrior', who is willing to protect civilians and dedicate his life for the good of the nation. This resonates with the expectations of the Swedish male conscript's role expressed in the law on compulsory conscription: "The soldier must be instilled with such a sense of *duty* that, even under a threat to his life, he shall comply and fulfill his duty without hesitation. This is the foundation whereupon ultimately lies the army's fighting capability in wartime" (1941 Defense Report, cited in SOU 1965:68, 34; in effect until 1990 [SOU 1990:89]; italics added). While the 'just warrior' is a historic imaginary, it is related to the ethics of war. The ethic involved in Swedish conscription practice is that of neutrality; the 'neutral soldier' acts in defense only, to protect the national territory and its people if attacked.

Because military and defense activities have historically been tied to processes of national identity formation and state-building, the 'just warrior' is also an important model for the citizen. It defines who the citizen is as well as his or her citizenship (i.e., the rights and obligations citizens have toward the state for the nation). As George Mosse (1990) shows, men have been the norm for military practices and for defense and security activities. Women, on the other hand, have been either excluded from these organizations and their activities or perceived as deviant. The association of the 'just warrior' with male citizens has historic roots; as Nancy Hartsock (1983, 186–209) reminds us, the male citizen became the norm in the Athenian polis, and he had a higher ranking in society than a female citizen. This was tied to his willingness to give his life for his country. Citizenship and nation-building hinge on the notion of sacrifice, and 'the just warrior' sacrifices his life for the nation. We can argue, as does Enloe (1993, 17), that any type of militarization demands that men be willing to die for their country while simultaneously requiring that women be willing to be the protected ones. Just as the protector needs something to protect for the relationship to be complete, there can be no masculinity without a complementary construction of femininity. Hence, gendered identities complement one another in the Swedish national defense.

The construction of male citizenship prescribed not only a specific role for the male—as the conscript, soldier, or military officer—but also a particular role for women outside the realm of the military, for example as mothers, nurturers, and caretakers. Women as the 'beautiful souls' have their given and specific role associated with femininity. In practice, women are expected to service, nurture, and nurse the citizen-soldier; reproduce new soldier-citizens; and take over the soldier's civil work in wartime while returning to the traditional role during times of peace (Kaplan 1994; Reardon 1996). Here, on a more symbolic level, the 'mother' image is particularly salient because the mother reproduces the nation biologically and culturally (Yuval-Davis 1997). The image of the female citizen as nurturer and caretaker has also been crucial to Sweden's national security and defense. It follows that, under neutrality, a woman unfit or unwilling to bear children was not a full citizen, nor was the man who refused to bear arms and take on the protector role.

As long as the gender dichotomy is perceived as given and natural, life proceeds as usual and little conflict arises. The association among men, masculinity, and military and defense activities represents one of the most historically cemented and universal understandings of citizenship (Goldstein 2001). In the Swedish context, such gender dichotomies have prevailed for a long time, despite the fact that military career positions have been open to women since 1980 and female conscription training has been voluntary since 1995. In interviews conducted in 2000, Erika Svedberg and I spoke with male conscripts about their role as conscripted men. Many young male conscripts indeed expressed doubts whether women have the stamina or the physical strength to fulfill the requirements of basic training. However, when asked about their own role as conscripts, many young men felt that they were naturally suited to defend the country and wanted to do what was asked of them. They commonly referred to tradition and conscription as a man's duty and thought it was natural to do military training because 'it is something one should do'. The 'taken for granted' feature of the gender dichotomy in Swedish conscription and military practice during neutrality was that men were 'naturally' soldiers, protectors, and conscripts.

Something perceived as natural and given is generally not thought to be in need of explanation or worth questioning. Often underlying what appears to be common sense are relations of power (Kronsell 2006). For example, a power relationship is embedded in the perceived-as-natural relationship between the protector (the conscripted soldier) and the protected (the nonconscripted). Cynthia Enloe (2007, 60) writes: "It is much easier to claim the authority to speak for others if one can claim to be The Protector; it is much easier to be silenced and to accept that silencing if one absorbs the self-identity of The Protected." Hence, the gender order of conscription was difficult to question, in the Swedish context particularly, because the dominant women's movement and feminist groups during neutrality were also antimilitarist. To question their status as the protected might mean they would also be expected to get involved in military training, and this would challenge their antimilitarist position.

The gender dichotomy was also relevant for Swedish nation-making outside the military and defense area. In labor and economic relations, citizenship was based on a social contract whereby worker and breadwinner became synonymous with the male citizen (Hirdman 1998). This established a clear hierarchy between men and women in the economic sector. 'Men's work' was paid work, largely in the private sector, while 'women's work' was unpaid work in the home or lesser-paid work in the public sector. Men as workers and breadwinners became given categories in the construction of a national identity, while women were excluded from these categories or perceived as a flexible workforce. For example, parental leave rights and support were organized based on the image of women as mothers. As pointed out earlier, a similar social contract materialized in conscription law: The male citizen was obliged to sacrifice his life for the nation and in return was given certain rights. The female citizen was sometimes encouraged to engage in the civil defense; however, any direct involvement on her part in the defense carried out by the Swedish Armed Forces (SAF) remained voluntary because she had other duties. This social contract seems to have been widely recognized.

As stated by Inger Segelström (quoted at the beginning of this chapter), women do their duty to the nation when they bear children. Segelström connects this to the gender dichotomy of the Swedish labor market and says that "if women were conscripted, then parental leave[1] would have to be compulsory for men." She connects the duties with specific rights and privileges characteristic of nation-building in *folkhemmet*.

Conscription is interesting in relation to nation-building because, as Mjøset and Van Holde (2002, 38) claim, conscription was a significant institutional innovation that linked military service to citizenship and also had a far reach. Few institutions of the Swedish state have touched on the lives of as many men as conscription has. Well over 8 million men have completed basic military training in Sweden since 1812 (Ericson 1999, 13). The system of conscription remained stable during the twentieth century and required that every male citizen be prepared to defend the country from the year he turned 18 until the age of 47. This is not unique for Sweden; conscription has in general applied only to males and has often coincided with their democratic rights (Sørensen 2000, 314).[2] Conscription practice thus gendered citizenship, with the man becoming the full or prioritized citizen. Following basic military training, men would return to their homes, dispersed throughout the country, where they were given a specific role in an eventual mobilization for war. Another option for men over 47 was to enroll in the Home Guard or in one of the voluntary defense organizations.

Margaret Levi (1997) shows that the willingness and consent to give one's life for the sake of the nation has been necessary to mobilize massive military forces but also that this consent was obtained only with some effort. Conscription was part of a bargaining process between the state and the male citizens, whereby the state offered citizenship rights to men in exchange for their willingness to sacrifice their life for the state and nation. It is easy to imagine that citizenship rights were perceived as an insufficient reward for the willingness to sacrifice one's life. In his study *War and Gender*, Joshua Goldstein (2001, 252) argues that gender became a tool with which societies induced men to fight. Soldiers require intense socialization and training to fight effectively; this is not a 'natural' or given trait of men, and making military service a natural part of men's lives requires significant effort and training. As Mosse (1990) argues, the making of men and establishing of masculinity have been extremely important to militaries both in wartime and during the preparation for war. This is also reflected in the words of former chief of the SAF Bengt Gustafsson (1995, 137) when he described conscription as important for its positive influence on individual male identity formation as well as on society at large: "The compulsory conscript system represents a value in itself contributing to stability in society. For most 20-year-old men, service in the SAF represents a valuable transition from the situation of the boy living with his parents to that of grown-up man integrated into working life." In this way, gender is an important resource in military practice. Conscription is constructed as a necessary component in the maturing process of a young man (i.e., it constructs manhood).

Masculinity (and the threat of not being masculine enough) has been widely used to motivate young men to participate in military service. Traditional rituals have been modified and incorporated into military training. Such manhood-making rituals encourage warrior qualities—these rituals use shame, encourage men to suppress their emotions, and, above all, differentiate warriors/men from women, for example by feminizing the enemy (Goldstein 2001; Jacobsson 1998; Mosse 1990, 61). Thus, notions of masculine identity and a privileged position vis-à-vis other citizens (such as women, elders, children, etc.) were reinforced and tagged on to other 'rewards', resulting in a gendered social contract between the state and the individual man. While taking on the role of the protector of the nation granted certain political rights of citizenship, it also provided additional benefits in terms of societal belonging as well as economic and social privileges (e.g., Wollinger 2000, 125–134).

MILITARISM AND CONSCRIPTION

Conscription training introduces the conscript to manhood while teaching militarism. Brian Fogarty (2000, 100–103) argues that military service is a very important site where norms related to militarism are conveyed to young men (and a few women). Conscription, because it involves such a large percentage of the male population, has a particularly influential part in this. Militarism is about defining the strength of the nation and its well-being "in terms of military preparedness, military action and the fostering of (or nostalgia for) military ideals" (Bacevich 2006, 2). For Ben-Eliezer (1998, 8), militarism is a cultural, even ideological, phenomenon; it is not a direct cause of war but makes war seem "a possible, available and reasonable solution." For a state to make war or defend itself against an aggressive enemy attack, a certain military preparedness must exist, with equipment, military forces, and so on, but it also requires "a culture of militarism; that is, military values and ideals must permeate civilian life" (Fogarty 2000, 84). How can this be relevant to a nation that has called itself neutral and has not been in a war for over 200 years?

Militarism does not necessarily imply an excessive use of military violence (Carlton 2001) but is a norm system deeply embedded in the culture of states and militaries. It has relevance for Sweden even though the country was neutral during the Cold War. Although neutrality is a defensive security strategy, it is not the same as pacifism. In Sweden, extensive military forces, a weapons industry, and a large voluntary and civil defense force back it up. Thus, it is possible that the conceptualization of the 'neutral soldier' that emerged from the specific context of the Swedish *folkhemmet* also was influenced by militarism. Although Sweden is in one sense highly influenced by militarism, it is a nation based on neutrality. The image of the defender of the nation is as strong and armed but also implies a passivity toward external military relations, with nonalliance policies as well as a certain internationalist outlook with involvement in mediation and peacekeeping.

The Swedish conscript soldier of the national defense can be conceived as embodying a specific 'neutral soldier' masculinity, different from other warrior masculinities and reflecting the values cherished in *folkhemmet*. Hutchings (2008) supports the relevance of this when she argues that changes in the nature of war, military practices, or foreign policy do not sever the connection between militarism and masculinity. Rather, they tend to complicate the picture of masculinity. Hutchings exemplifies this by saying that different masculinities are associated with different wars, from virtual war to guerilla war and war-lordism to peacekeeping missions. A different type of militarism—in the case of Sweden, a 'neutral' one—was tied closely to the defense of the territory and associated with a specific masculinity.

Michael Mann's (1987, 35) definition of militarism can be adjusted to be applicable to the Swedish context: Swedish militarism during neutrality constituted a set of attitudes and social practices that regarded total preparation for and social mobilization of a military in the defense of *folkhemmet* a normal and desirable state. It was consistent with Cynthia Enloe's (2007, 4) ideas of militarism as the belief that hierarchy, obedience, and the use of force was particularly effective in defending the nation in a dangerous world. Conscription was part and parcel of teaching young men militarism, but it was somewhat ambiguous. Conscription practice was broadly perceived as a way to minimize the risk that militarism would spread in society. There was a widely held belief that if citizens took part in military training, the defense sector would become more transparent and open to public control, therefore counteracting the risk that militarism would run amok and that the military organization would use violence against the population. This was assumed to work both ways. A strong connection between military and civilian life would make the defense strategies known among the population and generate wider public support (Körlof 2001; SOU 1965:68, 139; SOU 1984:71, 63–64), whereas a certain transparency contributed to societal control of the military apparatus. This argument has been articulated more generally (cf. Janowitz 1983; G. Ritter 2002). In this respect Sweden held a long-standing political consensus on the benefits of conscription over, for example, a professional army. Conscription allowed a form of civilian control of militarist values and the military organization. From a political point of view, civilian values and methods had to have the upper hand. Not only was there no real fear about militarism being widely spread through conscription training there was also a recognition that different tasks and responsibilities should be allotted and kept strictly to different spheres, civilian and military.

Two other widely articulated beneficial elements of conscription practice to nation-building include its integrative and informative functions. First, conscription practice presented the opportunity to unite men across societal distinctions and boundaries, such as class, profession, rural/urban background, and personal interest (SOU 1965:68, 136, 139; SOU 1984:71, 3–4; SOU 1992:139, 7; SOU 2002:21, 173). This had a homogenizing effect on nation-building. *Folkhemmet* was about the involvement of all members of society in a common project and reflected a common purpose of political institutions (Lagergren 1999, 184). Conscription was thought to include men from all segments of society in a way that disrupted class, regional, or other distinctions and encouraged tolerance among different groups of men to build community. The young men were exposed to one another, which helped them mature and grow into manhood. The general sense was that everyone had a chance to show his worth in the military service. A majority of the conscripts that we interviewed said that meeting fellow conscripts was one of the most important aspects of the overall experience of military service. Particularly important in this respect was the opportunity to meet young men whom they otherwise probably would not have encountered (i.e., from different parts of the country, with different social backgrounds, and with different interests).

The efforts at building *folkhemmet* via conscription did not explicitly address the bridging of ethnic boundaries and did not seem very concerned with the fact that only a few citizens with ethnic backgrounds other than Swedish went through conscription. The potential of ethnic integration through conscription emphasized by some scholars was not discussed in the Swedish context (e.g., Simonsen 2007). Anna Leander (2004, 574) calls the integrative ambitions of conscription practice one of the myths of conscription. According to Leander, it is a myth because it is "silent on the crucial questions of who is part of society and which social groups are integrated" and which ones are not. As argued previously, there were different expectations of citizens depending on gender, and still there was no word on this.

Second, conscription was connected to nation-making through the alleged information function. Via conscription practice, security and defense information was disseminated through educational efforts. This way a communication link between the state and the individual citizen was created. The link between personal identity and collective identity is "provided through the process of socialization of the individual into the behavioral norms of his or her society" (Hall 1999, 35). Tomas Denk's research (1999, 199) showed that the strategy seemed to work. Conscription socializes. The conscription experience also meant an awakening to politics for many conscripts. Körlöf (2001) says that the conscript experiences a learning process whereby he gains vital knowledge and understanding of the military and the defense and security of the country. The conscripts we interviewed confirmed this and said they had gained a greater understanding of Sweden's defense and security policies. Some became convinced of the necessity of a strong defense, which can be seen as evidence that they actually were subject to and affected by a culture of militarism. To others it meant the opposite. They realized how wasteful and futile the defense efforts were, perhaps giving the conscript an insight into the problems of militarism. It is clear that conscription gave these men a considerable insight in the military and defense sector's politics and organization. If conscription had this information function, then women clearly would not have the same kind of knowledge regarding the country's defense and security.

A broad political consensus held that the features described made conscription a democratic and democratizing activity. Little debate or criticism has occurred of the 'myth' of the democratic argument for military conscription. It remained a nonissue during the entire Cold War period. To find out why, we examined the gender dichotomy embedded in the national security imagery of Sweden, which implies that not only military activities would be male coded but also the knowledge and information about security and defense. This view seems to have been institutionalized broadly as it too became common-sense knowledge—natural and perceived as given. Hence, women's exclusion from conscription's information and educating function was apparently not something to be addressed or worry about. Only in early 2000 when the number of male conscripts was severely reduced did the problem regarding the dispersal of security and defense information to the citizens become a concern. None of the earlier governmental investigations on conscription viewed the lack of information about the defense as a problem (SOU 1965:68; SOU 1984:71; SOU 1992:139). The governmental investigation on conscription from 2000 was the first to point out that if only one-third of the men go through basic military training, an information gap is bound to develop, posing a potential threat to the country's safety (SOU 2000:21, chap. 7). This report concluded that efforts to spread information about Swedish security and defense policies became all the more important when so few were actually receiving military training (SOU 2000:21, 173ff). This 'information gap' becomes a real concern only when conscription duty involves less and less of the eligible male cohort, and thus large segments of the male population no longer "naturally" receive this information via conscription. The fact that women have always been excluded from information and knowledge on the nation's security and defense due to their exclusion from conscription and the military profession is of no concern at all.

This reflects a specific gendered citizenship. The male conscript's role in the nation in terms of the 'neutral soldier' extends well beyond the military institution. Men are perceived as actors in the public arena, and thus it is important that they are informed about the defense. Ruth Lister (1997, 70) puts this succinctly: "The citizen is the male individual who acts in the public sphere, representative of the household which he heads through the hierarchical institution of marriage, the locus of male citizens' power over female non-citizens." Conscripted men as heads-of-households may, if they wish, convey knowledge and experience of the public arena to the home and the private sphere. Hence, women either do not need to know about the military, defense, and security policies because they have a different role to fulfill toward the state, or because they are expected to gain this knowledge through their husbands, fathers, and brothers.

From a feminist perspective this is deeply problematic. In addition to either excluding women or treating them as 'others', the gendered character of conscription and military service has also denied women access, participation, and voice in defense and security matters. In other words, the problem is not only in the training of citizens, who are defending their country through a gendered conscription law, but also in the access to information about the defense and security of the nation. Citizenship and democracy has continued to be an issue for men but not for women. During neutrality the effects of this were substantial. Women's organizations, female politicians, and female citizens were largely absent from the debate on Sweden's defense. The gendered citizenship resulting from conscription has deprived women of agency in the military and defense debate. In addition, this may very well have excluded important perspectives on military and defense matters. Through opinion polls it is evident that in the area of foreign policy and security issues—such as Sweden's membership in the European Union and the support for a continued nonalignment policy versus Swedish NATO membership—there are considerable gaps between men and women's views (Stütz 2006, 2008). From a democratic point of view, it is troubling that these diverging views have not been included in the political debate due to women's lack of agency in security and defense matters.[3]

MASCULINITIES, FEMININITIES, AND
THE VOLUNTARY DEFENSE ORGANIZATIONS

Another area of defense activities important to building the national defense during the Cold War and as a part of nation-making and citizenship was the role of voluntary defense organizations. Many of these originated from the interwar years or earlier and were a vital part of the national defense and *folkhemmet*. Twenty-three different voluntary defense organizations were to provide additional personnel and skills, perceived necessary for the national defense. The voluntary defense organizations were tied to the national territorial defense organized throughout the country and in many cases to the self-sufficiency ambitions of Cold War neutrality. Over half a million citizens (around 8 percent of the population) were engaged in such organizations across the country, mainly divided according to the functions they fulfilled, such as driving large vehicles, organizing civil defense in neighborhoods, caring for farm animals, and acting as a reserve force in the sea defense.

The civil and voluntary defense of *folkhemmet* can, in Fogarty's view, be considered an extensive degree of militarization (see also McEnaney 2000). It reached out broadly to a large sector of society and supported and complemented what the SAF and the Home Guard could do in terms of defending the nation (see also Sundevall 2011). Well into the 1980s, the civil and voluntary defense had a budget that, per capita, well exceeded that of the superpowers' civilian defense budgets. Sweden's spending in this sector was matched only by Switzerland, another neutral country (Cronqvist 2008, 456–457). Tasks performed within the organizations were gender differentiated and associated with different masculinities and femininities. Gender was performed through such tasks but also challenged and recreated through organizational activities, while simultaneously the nation was made and confirmed.

Sjövärnskåren (the Voluntary Sea Corps) typifies one kind of voluntary defense organization active at the time.[4] It was first a branch of the military and later became a voluntary defense organization. Some other organizations, like *Sjövärnskåren*, were closely related to military activities; others were more connected to the civil sectors of society; and in some organizations that relationship changed over time. The term *voluntary defense organization*, although widely used, is highly misleading.[5] Many organizations emerged from civil society and were based on ideas about social movement activism and the importance of citizens' engagement in the reformation of politics that emerged in the beginning of the twentieth century. A broad voluntary civil engagement from various societal sectors was part of the broader neutral defense, and the voluntary defense organizations were an important part of *folkhemmet*. The related concept of *folkförankring* was frequently used and meant that ideas about the nation's defense should not only trickle down to the people but should engage them and become part of their lives, what in Laura McEnaney's (2000) words would be interpreted as where militarization meets everyday life. However, during the Cold War the organizations' existence came to depend almost entirely on state resources, and the social movement character of these organizations slowly vanished. Over the years, the term *voluntary* associated with *defense organizations* lost its connection with civil society, and the meaning was reinterpreted. "Voluntary" has come to stand for the opposite of duty, as in the duty to defend. This is an indication of the voluntary organizations' complementary character. They complement conscription and the Home Guard.

The familiar gender division relevant to conscription is also reflected in voluntary defense organizations. Men joined a voluntary defense organization because they were required to do so; women joined voluntarily. At age 47, all men who had completed military training could choose to join one of the voluntary defense organizations or the Home Guard, or continue to be assigned a specific position within the military organization in the case of war. The duty that men had toward the state through conscription was carried over to the voluntary organizations. Certainly, the men understood this. A FAK (Voluntary Automobile Corps) member stated in a 2007 interview: "Membership has become voluntary only now." Women would join under different conditions and for different reasons. For men, a voluntary defense organization was connected to the military branch in which they had served. Membership gave them an opportunity to use the skills they had acquired, such as driving heavy vehicles or acting as electronic engineers, or to pursue such interests further. Men's engagement in a voluntary defense organization was an affirmation of previous roles, tasks, and duties and also an affirmation of the masculinity associated with that context. Here the social function of the organization played a role in bringing together those with similar interests and experiences in military training and in society at large. The organizations also confirmed identities and the individual's place in the broader collective—the men of the nation.

When it materialized in the voluntary defense organizations, gendered citizenship also set the conditions for the organizations. First, organizations with mostly male members did not need to worry about resources or recruiting members. A steady flow of members was provided indirectly via conscription, and resources were provided for education of these members and for contracts. On the other hand, organizations with mainly women members had to put forth recruitment efforts. Neither their membership nor their resources were guaranteed. This affected the internal workings of the organization.

One of the largest of the voluntary defense organizations is the CFF (Voluntary Civilian Defense Corps), an organization that works with civil defense in the local context. During the latter part of the Cold War, it was the most gender-balanced national organization in Sweden, the result of conscious recruitment efforts that began in the 1980s. The CFF had been asked to develop what was called the home defense *hemskyddet*. This involved CFF members signing contracts that directed them to plan and organize civil defense action in a neighborhood. Women without children were targeted for recruitment and thought particularly useful in these efforts. They were not busy with duties as mothers and thus could be expected to serve the defense in other functions. The men in CFF were assigned leadership roles, so a gender hierarchy was quickly established. Men over age 47 were the leaders, and younger women acted as lower-level volunteers, taking orders and instructions from the older men. Although members of the CFF were both male and female, the mixed-gender character was only superficial. For a long time, tasks and duties in the organization were diversified according to gender hierarchies flowing from the military and conscription.

Erika Svedberg and I also analyzed the relationship between tasks and gender identities in eight of the voluntary defense organizations and noted how gender perceptions extended also into the performance of different tasks (Kronsell and Svedberg 2007). Four of the organizations that we studied were dominated by male membership, and each member was charged with a distinct task. For FAK, the task was to drive heavy vehicles to help move the troops across the country; in the FMCK (National Federation of Voluntary Motorcycle Corps), members drove motorcycles to aid the defense sector with quick deliveries, often in difficult and inaccessible terrain; and the FRO (Voluntary Radio Organization) worked with technical equipment to assist with surveillance and communication. By analyzing texts, photographs, and interviews and through observation, we found that different masculinities were expressed in the different organizations. In the literature, the difference regarding masculinities is usually related to social categories and less to function and tasks. We found that different masculinities related closely to the tasks performed, and because the organizations are distinguished by the tasks they perform, constructions of masculinity varied with driving trucks, working with marine ships, assisting in local civil defense duties, or working with communication equipment. Most members of these organizations were men over the age of 47, so we can assume they were settled in their masculine identity.

Concern about identity issues in these male-dominated organizations was articulated as anxiety about the identity of the organization in relation to other voluntary organizations, the SAF, and society in general. There was much talk about different types of clothes and symbols that could make the organization recognizable to the public. It appeared to be more about being recognized and perceived as part of a specific organization.

However, masculinity was not challenged, nor was it doubted; indeed, it seemed rather unproblematic. This does not necessarily imply that masculinity was irrelevant or unimportant, but it is, as Connell reminds us (1995, 212), the way the gender order often works. Masculinity is reproduced almost automatically when men go about things as they always have. It is assumed that the type of masculinity established through military conscription is carried over to the voluntary defense organizations and does not need to be particularly stressed or even discussed. Masculinity is taken for granted and made and remade through everyday performance.

For women, the voluntary defense organizations were much different. First, membership was voluntary. Two organizations accepted women only: *Lottorna* (Swedish Women's Voluntary Defense Organization) and the Women's Automobile Corps. *Lottorna* is the larger organization and also the one closest to the military sector; its role has been to provide auxiliary services to the military and the Home Guard. Many women who enlisted as officers when the military profession opened to them in 1980 were members of *Lottorna*. Over the years, their involvement in military activities has accentuated conceptualizations of gender and femininity in the Swedish defense sector. *Lottorna* launched a 'woman-in-arms' femininity concept, which rested uncomfortably with and even challenged the military gender dichotomy of the 'beautiful souls'. The following example from the early 1990s illustrates this.

During an exercise in collaboration with the Home Guard (at the time, all male) and *Lottorna*, the men of the Home Guard came equipped with the weapons that they normally keep in their homes. *Lottorna* members were asked to practice using wooden sticks. The women of *Lottorna* found this absurd and after the exercise demanded an explanation. The SAF responded that the reason why *Lottorna* could not be trusted to use real weapons during the exercise was because it feared that if *Lottorna* members did keep weapons in their homes, as the Home Guard men did, it would be a security threat. It was perplexing to *Lottorna* that the SAF based their security analysis on the idea that the women would have less control over weapons in their homes. The implication was that a *Lotta* had less control over her boyfriend or husband, who might easily find and use this weapon, hence the security risk. The Home Guard men, on the other hand, were simply assumed to be in full control of their weapons and their partners and thus presented no security threat. This example shows that the military gender dichotomy as expressed in the national vision of *folkhemmet* was relevant in the SAF's conceptualization of *Lottorna* and the Home Guard. This conceptualization by the SAF implied that women and men not only had different power positions in the home but that they also pose different kinds of security threats.

In terms of function and tasks, handling a weapon, and the propensity to use it is male coded (cf. Myrttinen 2003). Thus, it suggests that gender differentiation goes well beyond the body to include also the tools, means, and accessories used in accomplishing tasks. Thus, it seems the SAF normalized and naturalized the connection among men, violence, and military equipment. The example points to the relevance of Cynthia Cockburn's (2011) idea that the gender order and militarism are connected through the continuum of violence; here we see that exemplified by the gender relations around the possession and control of weapons. The thread of violence—via the gun—runs from military training with *Lottorna* and the Home Guard into the home and the private sphere.

Lottorna has tried actively to reconfigure notions of femininity, particularly by confirming women's agency through a 'woman-in-arms' concept. They have done so by strengthening women's confidence in themselves as independent, skilled people who are well aware of what is going on in the public and around defense, military, and security issues. This view represents women as willing to take on responsibility for the nation through defense and military activities, but it is contradictory to the conceptualization of men and women/masculinity and femininity in *folkhemmet*. Identity work seems to be a recurring theme in the history of *Lottorna*. It has been a struggle against both gender notions in the public eye and in terms of how that confirms and shapes members' own subjectivity. Again, the association with the performance of gender-coded tasks is relevant.

A historic and dominating imagery associated with *Lottorna* is related to an important function of the organization in the past, which has been to provide the food for the troops when they are out in the field. The iconography typical of a *Lotta* is stirring and serving big kettles of yellow pea soup. Cooking remains one of the tasks the organization performs, but it is far from the only one. Yet this image sticks. Cooking is a task that can be related to woman as nurturer, the one that provides food and nourishment, and it fits well with the gender dichotomy of the 'beautiful soul'. *Lottorna* has considered it important to challenge and extend this image. An example is a recruitment campaign in 2000 that had the explicit aim to show a different picture, one that associated a *Lotta* with excitement, drama, and adventure. One ad showed a *Lotta* hanging from a military helicopter about to rescue a person in the water. Another depicts a *Lotta* taking off her helmet and revealing her long, blond hair as she exits a military fighter plane. *Lottorna* as an organization is struggling with a problem that we return to later in this study: the contradiction around the imagery of the 'woman-in-arms'. In the nation-making process, *Lottorna* has often been pushed back to the stove—to cook pea soup—to a well-known nourishing femininity, which is more compatible with the conception of femininity associated with neutrality. In this struggle for a reconceptualization around a woman-in-arms identity, *Lottorna* has had minimal support from the women's movement, which has not perceived this organization as feminist.

Kvinnliga bilkåren (Women's Automobile Corps), the female equivalent of the FAK, is the other voluntary defense organization that has an explicit aim of engaging women. Both Women's Automobile Corps and the male equivalent the FAK, engage in the same tasks, namely, to drive heavy vehicles. The difference is that men can obtain a license for driving heavy vehicles, a basic requirement for membership in FAK, while under conscription service. Women, on the other hand, needed to obtain the license and the Women's Automobile Corps provided this opportunity. The tasks that *Kvinnliga bilkåren* engages with are masculine-coded tasks. Like *Lottorna*, this organization works in line with familiar strategies other women's organizations have used to improve women's confidence: through advancing skills and performing tasks defined as male (i.e., driving heavy vehicles) and by doing so through collective activities among women only. These are classic empowerment strategies for women. It is possible to imagine what it might mean for a woman's confidence to be

Figure 1.2: "You learn to dare" membership recruitment campaign. Photo by Sören Lund.

asked—under stress and time pressure—to deliver a heavy tank through a snowstorm in the desolate northern part of the country, and then to do this well. The conceptualization of the women of *Kvinnliga bilkåren* is far from the gender dichotomy of the 'beautiful soul'. In fact it challenges such a conception. It is rather about achieving independence, status, and even power through mastering skills and knowledge about a male-coded task: operating heavy vehicles.

In the narratives presented by *Kvinnliga bilkåren*, the focus is often on the idea of a member also being a mother; in other words: "You can be a mother *and* drive heavy vehicles." It seems important for the organization to accentuate motherhood. It is further exemplified in one of the more successful activities that have been conducted by *Kvinnliga bilkåren:* to arrange traffic security education in primary schools. *Kvinnliga bilkåren* aims to show that engaging in a male-coded task does not make her less of a woman. It also challenges the male-coded task of driving heavy vehicles, showing that a mother can do it too.

The third organization dominated by female members that we examined in our study is *Blå Stjärnan* (the Blue Star), an organization that initially assisted the military by taking care of wounded horses during the Finnish Winter War. During the Cold War, their task was to replace farmers active in the Home Guard. Self-sufficiency was an important aspect of neutrality and it was important to assure that farm production could continue even in times of war, occupation, or threat of war. Blue Star members would milk the cows, care for farm animals, and do other farm tasks. A similar process, as we have seen in *Lottorna* and *Kvinnliga bilkåren*, finds its equivalent also in the Blue Star. Blue Star, too, is in part geared toward improving women's agency by giving them (particularly city women) the opportunity to learn more about animals and farming and to put that into practice, improving women's confidence in the process. Beyond this, the gender identity of the Blue Star does not seem particularly contested or problematic. After all, caring and nursing, even if it is for animals, fits fairly well with the gender dichotomy of the 'beautiful soul'; in fact, historically, milking has been a female-coded task. After the Blue Star organization revolted in the 1960s and did away with their military leadership, it has had less to do with the SAF. They now work only indirectly to assist the SAF, by replacing Home Guard men who own dairy farms.

All of the voluntary defense organizations established their relationship to the state through a contract. As we have seen, conscription provided such a contractual link between the state and the individual man. Thus, the relationship with men was first and foremost through duty and conscription. Women, on the other hand, who had joined these organizations voluntarily, could either remain regular members or sign a contract with the state to do a specific task. Contracts between the state and the individual members were mediated via the voluntary defense organization. This contract sets out the obligations of and tasks for each individual and indicates where he or she should be placed in case of a mobilization of the defense. The organization is responsible for providing education, the exercises, and the social activities that made this possible. The following example from the Blue Star illustrates how this contract placed the individual with a specific responsibility toward the state for the nation. This system of contracts was also the way that the military and defense activities were organized into civilian life, in this sense 'militarizing' everyday life in a way that fit with the neutrality strategy.

An individual Blue Star member with a contract was assigned a specific dairy farm. She was required to "practice" milking and doing farm work a few times a year, usually when the farmer was away exercising with the Home Guard. Through the contract, certain obligations were assigned to the individual (i.e., to be responsible for milking at a specific farm), and she established her place in the defense of the nation with the task of providing food and securing self-sufficiency, both highly important to maintain neutrality, as it could not depend on other nations for provisions. Many Blue Star members said that they felt reassured by this—that, in case of war, they would know exactly where to go (and they could bring their children) and what they needed to do as their part in the defense. All of the voluntary organizations worked the same way. Through contracts, the state could assure that specific members would be placed in particular positions to do their duty in the defense of the nation, whether this was in close cooperation with the troops, on a dairy farm, or in the local neighborhood.

Apart from contracts and training in relation to these, the voluntary defense organizations had an important educational component. Taking courses was one of the requirements for signing contracts, and all contracted members had to take a comprehensive and standard course on Sweden's defense strategies. Just as with conscription practice, the voluntary defense organizations were given an important educational function: to teach the citizens about the defense and security in terms of threats, policies, and institutions. For women, who did not go through conscription, membership in the voluntary defense organizations became a way to learn about these issues. For the Swedish female population involved in one of the voluntary defense organizations, it reduced the gendered 'information gap' to some extent and provided them with knowledge, otherwise lacking, that could potentially give them more political agency in defense and security issues in general.

CONCLUDING COMMENTS

This chapter took a starting point in what many feminist scholars have argued about nation-building: that it is a highly gendered practice. This was confirmed in the Sweden case. Swedish nation-building around *folkhemmet* was gendered. In terms of the *folk*, men were the 'neutral soldiers' acting as protectors of the nation in the name of a specific militarism associated with neutrality, and women were the 'beautiful souls', the ones to be protected and thus defined outside both conscription and military and defense practice. Their agency and role was commonly associated with motherhood. It was also gendered in terms of *hemmet* (the home), with its close connection to the defense of the territory and the gendered-differentiated tasks and responsibilities.

Conscription practice and the voluntary defense organizations were analyzed to show in some detail how gender notions of citizenship were inscribed symbolically. Women and men as citizens of the Swedish nation were subject to very different expectations, placed and organized in segregated ways, and associated with different tasks as well as different levels of security threats. While gender dichotomies are symbolic expressions, this analysis shows how they came to be expressed in institutional practices and at an organizational level. This, in turn, affected individual subjectivities, making it difficult to imagine the soldier and full citizen as anything but a man and difficult for women who desired to associate with the defense and the military to engage in any tasks related to this sphere.

The 'just warriors' became, in the Swedish context, 'neutral soldiers', male citizens who had become political subjects through conscription practice, their agency deeply integrated into the home and the nation through *folkhemmet*, as well as via the information function. The idea that women do not belong in, have anything to do with, or have anything to say about or anything to offer the defense and military was prevailing. Because women were defined as outside the military organization, as 'beautiful souls', their role as mothers of the nation was verified in the defense organization. Any other role that women attempted to pursue posed a challenge to the defense organization with the 'woman-in-arms' identity being particularly problematic.

The gender identity associated with women through *folkhemmet* was especially contested in the activities carried out by some voluntary organizations with exclusive female membership. Women and femininity was organized by the state in military conscription and the voluntary organization and had become connected to their role as mothers and to a construction as 'beautiful souls' of the nation, excluded from military practice as well as from information on security and defense issues. This circumscribed women's political agency. Through the analysis of voluntary defense organizations, we noted that women were not satisfied with their position as 'beautiful souls' but indeed were willing to do their part for the nation. The voluntary defense organizations were an important arena to empower women in the field of defense and gave women some political agency. For *Lottorna*, it meant ambiguities around establishing themselves as full agents in defense activities as women-in-arms, and for *Kvinnliga bilkåren* it led to the possibilities of both acquiring agency in relation to driving heavy vehicles and experiencing anxiety about that activity's relation to motherhood.

The paradox of the findings is that while the neutral strategy was to remain nonaligned in peacetime, with the ambition to stay neutral in the eventuality of a war, the national defense was influenced by militarism and this militarism was widely dispersed throughout society. Ideas and practices about the protection of the nation trickled down throughout *folkhemmet* via conscription and voluntary defense organizations, making it every man's and every woman's concern. Conscription duty meant that every man learned to handle weapons and kill, and weapons were kept in the home; many women were engaged through the voluntary defense organizations, as were older men. Indeed, there was a kind of defensive and 'neutral' militarization that connected to everyday life such tasks as driving trucks, cooking pea soup, and milking cows.

In the next chapter we switch to an organizational perspective to analyze the defense sector. That chapter focuses particularly on the military organization of the SAF and the way it organizes gender and sexuality, exposing some of the challenges that the SAF faces in a postnational defense context when, for example, equal rights are the norm. In the third chapter we return to nation-building but discuss its second dimension: the foreign policy of Sweden in a postnational defense context.

CHAPTER 2

Gender, Sexuality, and Institutions of Hegemonic Masculinity

An important finding of the previous chapter was that the Swedish military organization has been part of a gendered defense that has constructed a specific national identity and militarism, demonstrated through conscription practice and the voluntary defense organizations. Through such practices, a gendered citizenship developed in terms of two national identity constructions: the 'neutral soldier' and the 'mother'. The military was an important institution in this scenario. This chapter conceptualizes the military as an institution of hegemonic masculinity and begins with an introduction of the concept (Kronsell 2005). Then, using institutional and organizational perspectives, I analyze practices that make visible norms of hegemonic masculinity embedded in the military institutions. Gender and sexuality in the military are explored in three ways.

This chapter first takes the perspective of female soldiers and inquires how norms of hegemonic masculinity shape the possibilities of a 'woman-in-arms' subjectivity in the context of military practices. Through an in-depth analysis of how gender has been constructed, transformed, and challenged within the Swedish military, I argue that gendered norms become apparent when women engage with the military. This occurs as the female soldier's subjectivity is formed against the norms of masculinity embedded in the organization. Second, gender is considered as an organizational resource. In institutions of hegemonic masculinity, sexuality plays a significant role and relates to the construction of both heterosexual masculinity and homosociality. The example here regards group cohesion, a key feature related to and perceived as crucial to military performance. Third, the chapter assesses the strategies promoted by the Swedish Armed Forces

(SAF) in recent years to deal with harassment, and discrimination, which have aimed to increase gender equity and tolerance for difference. These strategies are interesting for two reasons: They promote norms and ideas that seem to clash with the hegemonic masculinity embedded in the military institutions, and they allow us to investigate how gender and sexuality are promoted, understood, and resisted in this context. The strategies also promote a postnational military organization, whereby democratic norms and the rule of law implemented in the organization becomes a significant imperative, further addressed in the next chapter.

THE MILITARY AS AN INSTITUTION OF HEGEMONIC MASCULINITY

I have argued that the military organization, due to its connection with the defense and survival of the nation, has been more influential than other societal institutions. A military organization's task is to implement the state's monopoly on violence and thus is part of the continuum of violence related to gender and war (cf. Cockburn 2011). The influence of military institutions is both material and normative and herein lies its hegemonic potential: The military can use force, and the threat of force, and organize men by law, but it also requires a consensual understanding that this is the way things should be done.

Militaries are institutions that have largely been governed by men and have produced and recreated norms and practices associated with heterosexual masculinity, surprisingly consistent across both cultures and time (Goldstein 2001, 10–34). In most parts of the world, as we saw in the case of Sweden's national defense, the predominance of men in military institutions makes the link between the soldier and masculinity seem normal. The notion of masculinity that has been associated with military practices and war, I argue, can be considered hegemonic. The concept of hegemonic masculinity has been criticized by some, such as Jeff Hearn, who argues that it smoke-screens material aspects of power: the hegemony of men in patriarchy. He warns us: "The individual holders of power may be very different from those who represent hegemonic masculinity as a cultural ideal" (Hearn 2004, 57). Certainly he is right. There is a material dimension, as military institutions remain dominated by 'male bodies'. Connell (1998, 5) says, "Men's bodies do not determine the patterns of masculinity, but they are still of great importance to masculinity. Men's bodies are addressed, defined and disciplined and given outlets and pleasures by the gender order." Indeed, the bodies that have been disciplined in conscription practice and mass armies over the centuries, even as 'cannon fodder', are not the men vested with power. Yet, as we saw in the previous chapter, in relation to women these men have gained citizenship, political agency and rights that women have not, as well as becoming trained in using violence (cf. Pyke 1996, 532). Thus, institutions of hegemonic masculinity have not only been monopolized by men's bodies, but the norms of the military as an organization are defined on the basis of male bodies and masculine practices (Higate and Hopton 2005).

Another criticism is that the concept of hegemonic masculinity tends to fix a model across history. Hegemonic masculinity refers to a particular set of masculine norms and practices that have become dominant. In the Swedish national defense, it was associated with the 'neutral soldier'. Hegemonic masculinity is never explicit, yet it is the norm and it is normative (Connell and Messerschmidt 2005, 832). To become hegemonic, cultural norms must be supported by institutional power, that is, associated with powerful social institutions. Indeed, this is the case for military institutions in which certain norms of masculinity have been entrenched and institutionalized. Connell (1995, 213) argues that the military takes a particular position as the most important arena for the definition of hegemonic masculinity in the U.S. as well as the European context. This may make the concept seem more fixed than it is, and that indeed is not the point here. This book aspires to look at the development of a postnational defense under which such embedded norms are clearly challenged. Yet, after following defense and military policy for over a decade, I remain convinced that a specific masculinity has been the norm for the military and that it is hegemonic because of the way it is institutionalized and normalized.

While the hegemonic masculinity associated with militaries cannot be considered normal in a general sense, it certainly is normative for security, defense, and military practices. Hegemonic masculinity becomes some kind of model of "admired masculine conduct," which may be far from the lives of men but expresses "ideals, fantasies and desires" (Connell and Messerschmidt 2005, 838). Evidence of this in the Swedish context, a nation that has not conducted a war for over 200 years, is the extensive sales and market for war literature and the interest in war culture and heraldics.

Thus, by using the concept of institutions of hegemonic masculinity, I denote a particular interest in norms relating to gender and sexuality in the military. One way to address this would be to study different masculinities within the institutions and how they relate to a hegemonic masculinity. A comparative study of the organizational culture of the Swedish Army, the Navy, and the Air Force showed, for instance, considerable variation in the norms of masculinity within different subcultures, with the Army mirroring most closely the "combat, masculine-warrior paradigm" (Magnusson 1998, 24; see also Ivarsson and Berggren 2001, 15). Here I address institutions of hegemonic masculinity from the perspective of women who challenge masculine norms when they become soldiers.

A determining characteristic of institutions, historically dominated by men, is that gender as well as sexuality are largely silenced issues. In the military, silence relates to men, their gender, and their heterosexuality. Women have a gender and a sex; men do not. For example, men are soldiers, women are female soldiers, and they are all assumed to be heterosexual. Sexuality is crucial to the military organization, an important resource for military performance, and closely tied to a heterosexual masculinity as well as homosociality.

When gender becomes a topic on the agenda of an institution of hegemonic masculinity, it tends to be equated with women. Women are associated with gender and sexuality, not men, because men are the standards of normality (Connell 1995, 212) and thus hegemonic masculinity "naturalizes the everyday practices of gendered identities" (Peterson and True 1998, 21). This leads to the rather perplex result of "there is no gender but the feminine" (Butler 1990, 19). For the military organization, masculinity is not about gender—it is the norm. Understanding the role of gender and sexuality in institutions of hegemonic masculinity requires a questioning of 'silence' around what appear as normal categories.

Women's presence entails a particular challenge to militaries, and, in the past, women have been excluded from national military institutions in a majority of countries; women still remain a minority in all contemporary military organizations. For the same reasons, it is interesting to study the effects equality and gender policies have on a military organization, particularly as it reforms itself in response to the demands for a postnational defense.

The military's hegemonic status in society is weakening in many countries. The SAF is an example, but in most European Union (EU) member states, considerable cuts in defense spending and restructuring occurred after the Cold War, which also led to traditional notions relevant to militaries being questioned. Also, in peacekeeping operations—an increasingly important part of contemporary defense activities—there is a strong imperative to include gender aspects through UN SCR 1325 and 1820. Together with the 'new values'. this is an added pressure for military organizations to take gender issues seriously.

The personnel or 'manpower' needs of the SAF have been secured mainly through conscription (Ericson 1999). Young Swedish men's role as the country's defenders, protectors, and soldiers was simply a normal, and even natural, condition, and "conscription has always been justified, using vaguely articulated, largely intuitive appeals" (Leander 2004, 591). Hence, the relationship between masculinity and the Swedish military defense was also taken for granted and remained an unarticulated premise of national defense policy (Kronsell and Svedberg 2001, 2006). As argued in the previous chapter, conscription normalized the role of soldiering and protection of the nation-state as a man's duty. It was exclusive to male citizens until 1995 and compulsory for men until a parliamentary decision of June 2009 made conscription gender-neutral while at the same time suspending it.

In Sweden, there was never a heated public debate either about women becoming military officers or including them in conscription. The lack of public debate on conscription can be attributed to the political consensus on the value of the practice. Sweden has maintained conscription practice as the preferred way to provide 'manpower' for the defense for longer than the rest of Europe because of the role conscription has played in constructing the nation (Leander and Joenniemi 2006). Although the issue has not been raised very often, in the occasional debates that have occurred, the most common arguments against extending conscription to include women have been framed around security and economic concerns. During the Cold War, a truly universal conscription would undermine the legitimacy of Swedish neutrality, as it would double the recruitment, entail a substantial cost increase, and appear to be a step toward militarization (von Sydow 2001; see also Ds 2004:30; Proposition 2004/05:5; Proposition 2001/02:10).

Although the practice of male-only conscription was resilient, lasting until 2009, attempts to include women in the all-male military began in 1969. The Air Force wanted to allow recruitment of women to military positions due to lack of qualified personnel (Kryhl 1996; SOU 1977,26, 179) and reflected a common approach within male-dominated professions, namely, when there is a lack of 'manpower', recruitment of nontraditional labor becomes an option. Since 1981 women have been allowed to join the military if they are interested in a career as officer. While this meant that the military was no longer a career option exclusive to men, it corresponded to a functional need of the military organization rather than a concern about equal employment opportunities. It also constructed women as different from male-conscripted recruits:

Men had a duty to defend whereas, for women, military employment remained voluntary.

When the topic of women in the military was first introduced on the political agenda in the early 1970s, there was some debate and doubt about whether women's bodies were physically and physiologically apt for military service. Those who were skeptical about women as soldiers focused on assumed biological differences between men and women. Such arguments continue to be voiced (e.g., Carreiras and Kümmel 2008b). This argument carried little political weight in Sweden in the 1970s, and it was perceived as archaic to exclude women from a career in the military—a state sector—at the time when equality in working life was fairly salient on the political agenda. During the 30 years that have elapsed since women were first admitted to the military profession, despite various political, legislative, and organizational efforts to reform the military profession, surprisingly few women have been interested in engaging with it. One possible explanation is that the military has been slow and even reluctant to deal with the norms of masculinity that have become embedded in the institution through historic practices. Blomgren and Lind (1997) argue that because the SAF has failed to deal with the inclusion of women in the military as an organizational problem, there is a discrepancy between the policies of gender equality and military practice. They explain that due to the failure of the military organization to see that norms of masculinity are deeply embedded in organizational practices, gender mainstreaming, equality, and gender policies have remained abstract and difficult to understand and implement in everyday military practices.

WOMEN-IN-ARMS: A DELICATE BALANCING ACT

Although women's presence, as officers and as conscripts, has been minimal in numerical terms, their accounts of daily life in the military is helpful in seeing how its gendered norms operate. The contradiction between the 'woman-in-arms' and the norms of the 'neutral soldier' embedded in militarism becomes evident when this is challenged by the female soldier/officer in day-to-day practices. Conscription is perceived as a male sphere, which is challenged and even threatened by the presence of females. One of the first women trained for armored unit duty said:

> This thing about driving tanks is so macho. That we are allowed to be involved in combat crushes the men completely. . . . The men pretend that they are here to defend women and children. Then as we [the women] join them in the fight, the whole thing becomes meaningless to them. (*Värnpliktsnytt* 1991, 18)

Indeed, women's presence was perceived as disruptive, exemplified by the opinion of a marine soldier who was highly aware of 'male games' being played as part of the conscription training process. He was convinced that the mere appearance of a woman would make it impossible to continue 'playing' and said:

> [I am] convinced that everything would fall apart if women joined. It would be enough to see a woman to destroy the masculine games going on here, just the presence of women would break it all down (Jacobsson 1998, 175; my translation)

This sentiment is not limited to certain individual conscripts but is more general. The feeling of being an outsider is a reoccurring theme among women conscripts and officers (see. e.g., Atterling et al. 2001; Pettersson and Persson 2005). In a survey among conscripts in 1996, 50 percent answered that they did not think women were suitable for conscription (*Värnpliktsnytt* 1996, 21). A study conducted by the Swedish War Academy in the late 1990s qualified the findings somewhat and showed that there was acceptance and support for including women in the military on an abstract level. However, when more concrete issues were at stake, such as communication, officers' roles, education, and relationships between home and college life, support dwindled (Blomgren and Lind 1997), as it also did when competition became tougher (Ivarsson 2002, 33–35).

Women who work in fields dominated by men and masculine norms often find themselves in a paradoxical position, expressed in some women's fluctuation between denying their womanliness and becoming radicalized by the situation. In a comparative study of the lives of women in the U.S. military and the Catholic church, Mary Katzenstein (1998) showed that when women were busy with their daily tasks, they were also confronted with the gendered norms of the institutions. This 'discovery' made gender norms visible. When masculine norms completely dominate, as they do in military organizations, they can give rise to radicalization and resistance. Some women attempted to change the norms, as one study of female officers concluded that they became more critical, less accepting, or more 'radicalized' the longer they stayed in the military (Pettersson and Persson 2005, 15). That other women would choose to live with and accept the norms of the military is not surprising. If norms are hegemonic, they also rely on consent and participation by those excluded (Connell and Messerschidt 2005, 841). Indeed, many female soldiers say that adaptation and adjustment to the circumstances is a preferred strategy (Pettersson and Persson 2005, 13–14). Everyday routines and practices reproduce institutionalized norms. Those practices constitute a situation in which norms of hegemonic masculinity are reproduced while simultaneously challenged and possibly transformed.

Karen Davis (1997, 185) writes: "As women enter the 'man's world' they are struggling not only with questions surrounding their capability to do 'men's work' but also with issues surrounding their own identity as women." Something similar happens here, when women cope with and try to adapt to the military organization, they are confronted with norms of masculinity but also norms of femininity. Women who take part in the military as conscripts or officers are marked as different because men are the 'real' soldiers, and that is what would be expected from an institution of hegemonic masculinity. At the same time, the women are faced with norms about femininity or what it means to be a woman because, as argued earlier, gender constructions are relational (cf. Kvande 1999, 306). Hegemonic masculinity contains within it constructions of femininity that come to the fore in this encounter. The 'women-in-arms' femininity that female conscripts and officers embody is a challenge to the hegemonic masculinity of militaries, but it is not void of femininity; rather, as Cynthia Enloe (1993, 17–20) reminds us, masculinity as relational needs a complementary construction of femininity. Empirical evidence of this in the Swedish military shows that femininity is often described as a negation (i.e., what is not feminine). This does not mean that femininity constructions are absent from the military. On the contrary, femininity has been there all along, but it has been associated with the protected and with femininities related to nurturing, supporting, and caring. Such femininities can more easily be associated with civilian employees, who traditionally have been present alongside and often in military camps, kitchens, laundries, snack bars, and offices.

Consequently, a female soldier's subjectivity can be characterized as a delicate balancing act in reaction to the formal and informal yet institutionalized norms of the military institution. Helena Carreiras' (2008, 177) research on women in the Dutch and Portuguese militaries leads to a similar conclusion. Women must constantly manage their identity, often with much ambivalence (see also Sasson-Levy 2003). Constructing a 'woman-in-arms' femininity seems hardly possible, and female recruits struggle with this. The female soldier constantly has to manage her femininity. This is clearly a challenge because "girls who do voluntary conscription are far from any womanly stereotypical ideals. They are tougher," said two male conscripts (*Värnpliktsnytt* 1989, 9). Yet if the woman is too tough and perhaps too manly, as we later see, she is not feminine enough and can pose a threat to the men (cf. Gutek 1989, 65). On the other hand, if a woman is perceived as too feminine, she may be viewed as too sexual and too weak.

An extreme in either direction leads to problems. Female soldiers must juggle their femininity, avoiding or moderating some types of femininities that are disliked.

One such femininity that female soldiers and officers often distance themselves from is that of 'the bimbo'. Suddenly, the sex object as promoted in pornography and pin-up pictures is standing among the men. 'The bimbo' may disrupt and disturb the troops. It is threatening to an institution of hegemonic masculinity because, as Ruth Lister (1997, 70) writes, it "is the very identification of women with the body, nature and sexuality" that is "feared as a threat to the political order." The male soldiers might compete over the attention of 'the bimbo', or perhaps think she has no place among them. 'The bimbo' evokes conspicuous sexual desire, and when the object of desire—normally represented by the pin-up girl— stands beside the male soldier as a woman-in-arms, the norms become visible through the ensuing awkwardness.

'The bimbo' might also become an object of sexual harassment that may be a component in the masculine games that support male bonding and heterosexuality. The reaction to 'the bimbo' shows that masculine norms in the military are entangled with notions of women as objects of sexual desire but also always as 'others' outside the realm of military activities. For these reasons, to be accepted, female soldiers must distance themselves from sexualized femininities. This was pointed out in an interview with a female captain employed in the SAF since 1981. She said that one of the essential survival skills for women is to be very modest, to desexualize themselves both in behavior and clothing, and, for example, to never dress in a short skirt and high heels. Carreiras (2008) discusses this in her work. The control of women's sexuality seems to be "particularly amplified in the military environment," and "women's sexual behavior is a matter of organizational anxiety" (Carreiras 2008, 170) and is not an acceptable 'women-in-arms' subjectivity. Other narratives suggest that female recruits are expected to resolve this sexual anxiety, or at least moderate it, by their presence (cf. Eduards 2011).

The Swedish military speaks with pride about changes taking place when one or a few women join. When a woman joins a garrison, it often results in a 'shaping up' both in the language used among conscripts and the photos and magazines that are displayed in the barracks. The frequent use of sexist language and pornography in the military is widely known. Military training in Sweden, as in most other places, has often relied on abusive language that is sexualized and often associated with aggression and violence (Berggren and Ivarsson 2002; Jacobsson 1998; Meola 1997). This overemphasis on sexuality and pornography has been expected of young men but is also somewhat of an embarrassment to the military. Thus, the 'shaping up' of the language and the removal of pornographic pictures from soldiers' quarters observed to follow with women's inclusion in military institutions has been considered positive by both officers and conscripts. Whether such gestures originate from the soldiers or as a response to the command of a superior, such changed practices associated with women's entry into the military are welcomed. They are considered positive not only for the individual female conscript or officer but also for the military's reputation among other state institutions.

The femininity that is expressed in this 'shaping up' role resembles that of 'the mother' or 'the big sister', who disciplines the young boys to behave. This subjectivity is suggested also in accounts of how women make the military a more human place with a nicer atmosphere. One female recruit said, "If we are out on an exercise and I see that a guy is tired, I will give him a part of my chocolate. It is some kind of mother instinct, and I don't think a guy would do that" (*Värnpliktsnytt* 1995, 16). Some claims that women are more psychologically stable, and finding a feminine identity as 'the mother' might provide a safer haven, as it is also a desexualized identity. Yet, as discussed in the previous chapter, 'the mother' is connected with motherhood, to be protected by the male soldiers, the protectors.

A problematic femininity in the military is the 'weak girl'. One female conscripted soldier said, "It is possible to retain your femininity in the SAF; you don't have to turn into a tomboy, but you definitively cannot be girlish" (*Värnpliktsnytt* 1989, 9). The 'weak girl' is to be defended; she needs protection. Her presence appeals to the kind of heroism and chivalry that has historically been associated with the male soldier (Kümmel 2008; Riemer 1998), and thus she cannot be a soldier. The weak girl-in-arms is a contradiction in terms, and the soldiers have to treat her differently—carry her backpack and help her through tough exercises. In this way, she violates the norm that all soldiers are equal. The requirement of not being too weak has both a physical and a psychological dimension. It is not only about being able to carry a heavy load but also about the ability to push oneself, to take one's share of responsibility in a group, and to take on a challenge.

The 'weak girl' femininity suggests that favoritism toward female recruits may exist. Female recruits often dislike paternalist behavior, that is, when they get help even when they do not ask for it. Kümmel (2008, 195) calls this type of "chivalry" "benevolent sexism." Most female soldiers want to show their skills and worth in the military. This has actually been an important motivation for women who have volunteered for conscription in Sweden (Atterling et al. 2001; Bjelanovic 2004). Female soldiers are afraid that positive discrimination will lead to envy and disrespect from the other recruits (Carreiras 2008, 171; Moelker and Bosch 2008, 112). Pettersson (2008) investigated how female cadets in the officers' program at Karlberg Military College perceived positive discrimination and showed that they were ambivalent. Out of the 16 women interviewed, 9 favored positive discrimination because they felt it was necessary for the proportion of women among officers to increase. Five were negative because of what it would imply for their daily practice in officer training. They were afraid that their male colleagues would not respect their competence and role as officers if they were favored.

A more recently articulated femininity, also undesirable, is 'the feminist' as a woman-in-arms. This includes women who actively pursue gender issues in a different context, where they discuss these issues and too eagerly point out the use of sexist jokes or language. Women officers and conscripts do not want the label of 'feminist'. They perceive it as very negative and shun it (Pettersson 2008, 8). The widespread contempt for 'the feminist' is suggested by examples whereby female recruits choose to make a point of not being feminist by loudly criticizing or ridiculing lessons on gender and equality, that way marking that she does not belong to the 'feminist' category (Pettersson and Persson 2005, 16). 'The feminist' is related to the context of the SAF that today is more actively pursuing gender equity issues.

Many narratives have indicated that 'the manly woman' is also a problematic woman-in-arms identity widely rejected by recruitment officials, conscripts, officers, and other women, albeit not in official sources. During the Cold War, female officers dreaded turning out like a "female Russian shot putter" (Nilsson 1990, 11). Female shot putters evoke the image of an androgynous, manly looking woman, big and strong, hairy and unattractive, with dysfunctional hormones (cf. Dowling 2000, 198ff). 'The manly woman' is not unlike the shot putter; she too has taken on too many traits of the masculine world, not only in physical appearance but also in inappropriate female behavior, like drinking excessively, being rough, and swearing (*Värnpliktsnytt* 1995, 18; 1998, 13). Women's inferior physical strength is a classic argument (Carreiras and Kümmel 2008) evoked against women's suitability to become soldiers (cf. Maninger 2008; van Creveld 2001). Thus, in a way it seems logical that 'the manly woman' would be most suited to serve the military. She would more likely be fit for the physical challenge, and her masculine behavior would fit easily into a military unit. This has not been the case in the Swedish military.

On the contrary, 'the manly woman' is a despised femininity for a female officer or conscripted soldier. Both male and female soldiers have articulated this quite strongly. 'The manly woman' can compete in strength, toughness, and roughness with men, and her unfeminine, almost masculine presence challenges the gender dichotomy. So even if 'the manly woman' evokes a woman who is more like a man and thus more like the normal soldier, the blurring of the boundaries threatens clear distinctions around gender and sexuality associated with hegemonic masculinity in the military. Particularly by evoking homosexuality, 'the manly woman' challenges the heterosexual norms of military masculinity.

Women soldiers express concern that they might become too masculine in military training. At the same time, it is clear that bodily strength, endurance, and physical achievement are necessary for a woman to fulfill her soldier role. This is a necessary feature of an acceptable 'woman-in-arms' femininity, while at the same time it cannot be overexaggerated in its bodily or behavioral expressions to the point that it resembles 'the manly woman'. Thus, for the individual female soldier or officer, the balancing act involves being prepared for the physical challenge and the rougher masculine comradeship while not being perceived as manly or masculine. In the end, the femininity associated with the 'woman-in-arms' remains highly volatile. Women choose to be both same and different. Thus the goal is to be 'one of the boys' although not identical to a boy or man. To fit in, a woman may go to great lengths to be perceived as 'one of the guys', and will do whatever she can to be seen as a soldier, just like all the men. However, it is important that she does not go too far, to become manly, because then she might be disciplined.

Carreiras (2008, 171) also found that women soldiers are often disciplined with special requests regarding their physical appearance, for example, asked to shave their legs or to not to shave their head. In a report on sexual harassment practices in the SAF, 40 percent of female officers and conscripts reported that they had experienced harassment in terms of male colleagues voicing opinions about their behavior. Interestingly, the attempts by male collegues to 'discipline' the female soldiers were made on the grounds that they were either too aggressive and provocative or too soft and careful. Both types of complaints were equally frequent (Ivarsson et al. 2006, 3ff). Thus being a woman-in-arms is indeed a difficult balancing act.

SEXUALITY IN INSTITUTIONS OF HEGEMONIC MASCULINITY

Sexuality plays a crucial role in constituting gender relations. In Catharine MacKinnon's (1989, 131) words, "the ruling norms of sexual attraction and expression are fused with gender identity formation and affirmation, such that sexuality equals heterosexuality equals the sexuality of (male) dominance and (female) submission." In other words, as stated by Hartsock (1983, 156), "sexuality must be understood as a series of cultural and social practices and meanings that both structure and are in turn structured by social relations more generally." Sexuality has been analyzed in organization and management studies as an element of the production of gender relations in society, and sexuality is considered key in organizational processes (Hearn 2010; Hearn and Parkin 1995). Sexuality has been incorporated in the organization of factories, offices, schools, and hospitals (Burrell and Hearn 1989, 12). Foucault's (1979) genealogies on sexuality also showed this. Sexuality as an expression of gender relations has become part of the basic assumptions of the institutions, that is, the normality of organizations and institutional practices.

With Foucault's (1979) view of power in mind, Peter Fleming (2007, 240) argues that organizational sexuality represents both domination and resistance and indeed is expressed in the organization in various ways, such as in sites of control, power, and sometimes violence, but also as empowerment and resistance. This is relevant for institutions of hegemonic masculinity because, as argued about militarism previously, sexuality, power, and violence are interconnected in the institutions of war and defense. These interconnections have been institutionalized as powerful norms that have informed military practice (Hearn and Parkin 2001, 13–14; see also Münkler 2005, 81–85). The power of sexuality indeed has been crucial to military operations in general (Goldstein 2001, 251–379).

Nancy Hartsock's (1983) notion of *eros* can be useful in understanding how sexuality is relevant to military institutions. She argues that *eros* comes in two forms: the body and the soul. In militaries, the two forms are closely connected, for example, in the practices around military cohesion. Hartsock (1983, 195) explains that one form of *eros* "has its place in politics" and "rejects the body in favor of the soul." It is what builds community and bonds people together, for example in conscription training, as buddies, and as partners in different types of social and organizational networks. This type of *eros* creates loyalties, friendships, and trust. While trust and community building takes place more generally among various groups, the historically dominant position of men in politics and public life makes homosocial relations among men the relevant focus. *Eros* is a powerful aspect of human relations, and thus it is not surprising that, as organizational theorists have argued, organizations and their leaders try to tap into it in different ways, usually by controlling it (Hearn et al. 1989; Sinclair 2005). We can thus expect that forms of *eros* are part of any organization and are tapped and controlled in different ways according to the organization's tasks or aims.

The way that *eros* seems to have been managed in military organizations is twofold: by encouraging, even overemphasizing, heterosexual sex, as long as its expression is directed outside military practices, and making sure that there is no sexual relations between the men within the military. The military prefers to enlist heterosexual men. In this way, *eros* is tapped to obtain homosociality, to be used against the enemy in troop cohesion, believed to enhance military performance. The two forms of *eros* seem to be interrelated. The bodily form of *eros*—sexuality—relates to the power of hegemonic masculinity in militaries.

Historically, pornography and even prostitution and rape have been (un) officially sanctioned by militaries and can be seen as a way to encourage, assure, and control heterosexuality among the enlisted. Cynthia Enloe (1989, 1993, 2007) has written extensively on how U.S. militaries officially sanction prostitution around military bases and that this has been part of the conscious activities of the U.S. military. In Sweden, much evidence shows that sexually abusive and misogynist language are used as part of military training and education. An example is the 'fighting songs' intended to get the troops in a fighting spirit; they are full of abusive sexual innuendoes intended to make the men 'fired up' in relation to the perceived enemy, which is feminized and sexualized (Jacobsson 1998, 95). Sex and sexualized violence have been important ingredients in the building of the military's and the soldier's identity (cf. Davis 1997, 185–186; Meola 1997) and thus integral to the organization. Hence, it is not surprising that sex, in terms of sexual harassment, is turned into a weapon against women in the military.

Studies have shown that when women enter areas in which men are in a majority, "men may use sexuality to maintain their dominant position" (Collinson and Collinson 1989, 103; DiTomaso 1989, 71). Sexist language and sexual harassment can be a way for individual men to assert their position of power over individual female officers or recruits who challenge the masculine norms of the military organization by being present and by engaging in male-coded tasks, such as handling a weapon. Gruber (1998, 314) argues that in organizations that are predominantly male, men "tend to mark their work environment with sexually objectifying material," and women are more likely to be harassed.

Sexual harassment against women remains a problem in the SAF, both in officers' and basic training. It was observed for the first time in 1989. A study conducted by the Swedish Defense College showed that the frequency of sexual harassment increased with rank (Blomgren and Lind 1997, 18; Ottosson 1997). This was verified in the SAF's survey in 1999 showing that 43 percent of conscripts and 59 percent of female officers said they had experienced some form of sexual harassment. A later survey on sexual harassment that included women from all categories (conscripts, officer trainees, officers, and civilian employees) verified that the more competitive the situation, the more common the sexual harassment (Berggren and Ivarsson 2002). Sexual harassment is used as a means to both affirm masculinity and protect or advance one's status in relation to women (Berdahl 2007). Sexuality becomes a way to protect a social status. A 2005 survey showed some improvement, 36 percent reported sexual harassment and the figure was the same for both conscripted soldiers and officers (Johansson et al. 2009).

A qualitative analysis of the survey answers (Ivarsson et al. 2006, 16) that compared the results from 2002 and 2005 argued that harassment of a sexual nature was more frequent in 2002 but that in 2005 the methods of harassment had diversified. The 2005 survey included three different types of harassment: (a) harassment with sexual connotations, such as sexist jokes, touches, and invitations that were perceived as disrespectful; (b) harassment due to the soldier's sex, such as being denied information or promotion or not allowed to do certain tasks, questioning her competence, or ignoring her; and (c) harassment related to provision of equipment, clothes, and sanitary facilities, perceived by the female soldier as creating obstacles for her and excluding her as a part of the organization (Ivarsson et al. 2006, 17). The different types of sexual harassment that the female soldiers perceived and reported in the survey relate to different levels: the interpersonal level, in relation to the leadership, and in relation to the organization's buildings, physical equipment, and so on. Understanding harassment in all these categories also means that harassment is potentially always there— in the way men look at and assign tasks to the female soldier, how she dresses or uses a gun, and the conditions of and how she is treated in the barracks. The study interprets sexual harassment and sexuality as multifaceted and working at different levels of the organization, confirming that organizational sexuality is not exclusive to personal relationships but is a broader organizational phenomena.

Sexual harassment is connected to the role of hegemonic masculinity in military institutions because "sexual harassment that women experience may be seen as a manifestation of the psychosocial pressures on men to identify with a form of 'masculinity'" (Thomas 1997, 135). Thomas stresses the link between the use of sexuality as a means of power and the need to demonstrate masculinity, suggesting that social norms dictate what is expected of masculine behavior (Thomas 1997, 145). Conforming to the norm of a soldier in the military, based on a certain idea of masculinity as the protector, requires a distancing from the protected or that which is feminine. Thus, conformity with the norms of masculinity embedded in the military organization requires a differentiation from anything feminine and a clear demarcation of heterosexuality. This can be achieved through acts of sexual harassment.

Epstein (1997) wrote that it appears that harassment directed toward women and gay and/or 'effeminate' men is also a way to enforce heterosexuality. For men, avoiding stigmatization and being able to communicate an acceptable masculinity seems to partly depend on harassment. Thus sexual harassment is implicated in the process of constructing heterosexual gendered identities (Epstein 1997, 167). Sexual harassment is not an individual strategy but is related to the institution of hegemonic masculinity in that it is a way to strengthen masculine subjectivity (Berggren and Ivarsson 2002, 61). Lees (1986, 31) remarks that such acts can be a means of promoting solidarity among "the lads," and she notes that sexism "appears to be one feature of male bonding, where denigrating girls and women builds up a kind of camaraderie between them." Such acts can combine the two elements of *eros*: sexuality and homosociality.

MILITARY COHESION AND HOMOSOCIALITY

There is more to *eros* than sexuality; the soul form is also highly relevant to understand gender relations. The interconnection between the forms of *eros*, or heterosexuality and homosociality, is particularly interesting in the military organization. The tensions between the two are related to the glorification of love and comradeship between men in war and military activities which gives rise to an anxiety whether it is the right kind of love. Public disapproval of homosexuality has been the way to make that clear to all.

This tension can be illustrated with a focus on troop or unit cohesion, apparently a central component of any Western military organization. In their classic study of the German Wehrmacht, Shiels and Janowitz (1948) argue that the integrative relationships in a smaller group is essential for the military unit to carry out its tasks under stress. Military scholars and practitioners continue to focus on the relationship between small-unit cohesion and military performance. Cohesion is generally understood as key to military performance and success. The discussion about the characteristic of groups and what this means for performance is a general concern, for example, in management studies and is relevant for most organizations, whether private or public. However, the discussion in the military has remained specific, most likely due to the alleged special nature of the military and the need to perform for the demands of war.

Recent discussions on cohesion among military scholars concern mainly how relationships become integrative and how to contribute to group cohesion in a way that increases military performance (e.g., Griffith 2007; King 2006; Kirke 2009). Military organizations seem to be the primary users of this knowledge, and they use the concept frequently. James Griffith (2007, 138) argues that "while there are many factors that motivate soldiers to fight, the nature of relationships within the small unit or group cohesiveness is one of the primary explanations in military literature." Carreiras and Kümmel (2008, 30) suggest that there has been a "renaissance of direct ground combat in military operation thinking" and unit cohesion has become an even more important issue in the war on terror.

The emphasis on unit cohesion in military institutions has also meant that homosocial relations have been privileged. The historic segregation of women and men in private versus public life formed public institutions according to a homosocial type of logic. The concept of homosociality means the seeking, enjoyment, and/or preference for the company of the same sex (originating in the work of Lipman-Blumen 1976; see also Kronsell 2009). Homosociality is about social ties and is a phenomenon that is subtle and intangible. While men might not consciously or overtly exclude women, they tend to be homosocial and thus favor connections with other men in work, research, and leisure time. Thus, men tend to reproduce the homosocial environment that they are part of and make it difficult for women to enter the scene (Bird 1996; Flood 2008). Men's homosociality is of particular concern in institutions of hegemonic masculinity, as is its role in conserving gender relations.

In institutions of hegemonic masculinity, homosociality is important, resilient, and powerful. Homosociality seems to have been tapped into to create masculine comradeship, a powerful positive force in unit cohesion (L. Skjelsbaek 2001). As Siebold (2007, 287) writes, military group cohesion is about both building trust in the smaller military unit (like a platoon) and the relation to the broader military organization. It has both an instrumental and task-oriented aspect and an affective dimension. Homosociality relates particularly to the affective dimension of cohesion. Cohesion has been perceived by many military strategists as the glue of the military and the basis for combat readiness but also as possible only between heterosexual men (cf. Magnusson 1998, 25). Masculine rituals have had a key part in creating and sustaining cohesion (King 2007, 638; Siebold 2007). The historic and systematic exclusion of women and homosexuals from the military can be seen as a way to ensure the development of homosociality, encourage masculine comradeship, and assure combat readiness. The military literature presents various perspectives on the relationship between group cohesion and military performance. For my purposes, it suffices to know that it is widely accepted that bonds that tie individuals in the group together determine whether they can perform their designated tasks well under extreme conditions and stress and whether they are able to perform in the interest of the group, for example, sustaining fighting to assure the survival of the other group members (Griffith 2007, 140).

Cohesion is a highly interesting aspect of military institutions because it aptly illustrates how *eros* is a resource to the organization and its operation based on the relation between heterosexuality and homosociality. Because cohesion has been considered important to military performance, it has been used to argue against women's and lesbian, gay, bisexual, and transgender (LGBT) persons' inclusion in the military. This has raised debates, particularly in the U.S. context (Barkawi et al. 1999; Kier 1998; MacDoun et al. 2006), where for a long time the military remained committed to this view on cohesion and excluded LGBT persons from military service by endorsing the 'don't ask, don't tell' policy on homosexuality. Some military strategists continue to argue that women and homosexuals are a threat to unit cohesion and compromise military readiness (Knapp 2008). U.S. President Barak Obama's commitment to repeal the policy came with legal and political controversies (Burrelli 2010) but succeeded in December 2010. Republicans and parts of the U.S. military establishment articulated views against the repeal, based on this potential threat to cohesion and performance ("Repeal of 'Don't Ask, Don't Tell'" 2010; "U.S. Judge Orders an End to 'Don't Ask, Don't Tell'" 2010).

Because women, on average, are physically weaker than men, some argue that women cannot carry their weight in a combat situation (Maninger 2008). Cohesion is threatened by the presence of a woman in the unit because men, it is assumed, will act chivalrously and attempt to protect her. Hence, unit cohesion depends on the individual being ready to sacrifice himself for the entire unit. The presence of a woman or a homosexual challenges this. As such, unit cohesion is about the relationship of masculinity and femininity, homosociality and homosexuality, and what is expected of men in a woman's presence. The threat to cohesion is more about what happens to the group when a woman or a homosexual man or woman is present than about that individual per se. The female soldier or combatant becomes both a distraction and a temptation and cannot be trusted to generate optimal military performance. Thus, women's presence becomes a problem for the bonding between men in the group (Jung Fiala 2008, 56). This ties in interestingly with what Ruth Lister (1997, 70–73) writes about women as a threat to the political order because of the way they are identified with the body and with sexuality. It is assumed that the woman or the homosexual man or woman is the threat and that nothing is wrong with the group or the organization. This idea also clearly builds on the notion that men are the protectors and the chivalrous ones (cf. Kümmel 2008), willing to sacrifice themselves for a woman or one another. Women, constructed as the protected ones, cannot be trusted to do so. Homosexual men or women also cannot be trusted, but what is at stake is homosociality.

This gives rise to real military policies, for example, by which homosexuals are excluded from the military profession and women are not allowed to serve in combat roles. As pointed out previously, the U.S. military has in the past made such restrictions on both homosexuals and women. Many NATO countries have similar restrictions on the presence of women in combat positions ("Percentages of Female Soldiers in NATO Countries' Armed Forces" 2007). The Dutch, for example, exclude women from the submarine service and the Marine Corps (Moelker and Bosch 2008, 107), as does the UK military (Woodward and Winter 2004, 295). To exclude women from certain military posts and career paths within the EU actually violates the EU Equal Treatment Act. The Swedish military does not exclude women from specific posts, as discussed in the study preceding the legislation that allowed women into the military profession (SOU 1977:26). In this Sweden seems rather unique. However, for LGBT issues, anecdotal evidence indicates discrepancies between formal rules and practices in the SAF; during conscription tests men were carefully screened for homosexuality and often dismissed because of it.

In the United States, where there is a limit on women in combat positions and have been restrictions on homosexuality, this issue has been researched and discussed extensively. Lieutenant Colonel Baker (2006, 5–9), for example, writes on the restriction on women in combat and argues that this has no base in research. Baker refers to studies that show that women do not negatively affect the morale of a unit or its readiness. Baker argues that it is the attitude of soldiers that determines cohesiveness, and hazing, harassment, and other conduct negatively related to homosociality can be detrimental to true unit cohesiveness (Baker 2006, 5; see also Kier 1998). This appears to resonate with the experience and view of General Karl Engelbrektsson from the SAF. In an interview in 2008, he talked about problems caused by internal conflict in military units and the damage sexual harassment can do to unit cohesion. He first observed this on a Kosovo peacekeeping mission (2003–2004), which was the first time he realized why and how gender issues could have relevance for the operational context of the military. To him, gender became equated with problems of harassment and bad behavior and affected unit cohesion. He explained:

> The task is to be ready, in unknown territory with difficult conditions. If, in such a group, there is no trust between the members under normal conditions, how

can it possibly work, when it is dark, difficult and dangerous? To eliminate any form of harassment becomes a matter of Force protection, that is, you protect the unit by making sure the attitude and behavior is conducive to building trust.

The Swedish General endorses the importance of group cohesion for military performance and sees trust as a crucial issue for cohesion. However, his view of cohesion is different, as he takes for granted a more inclusive military organization from the start. The issue is to deal with harassment practices in the entire organization in order to create the right trust conditions to optimize the unit's cohesion. To take a gender perspective is, to him, one way to do this. His view reflects the context of the SAF by arguing that intolerance, abuse, and bad behavior among those who are part of the force endangers the entire group. This shows cohesion can be interpreted and understood in alternative ways, and not all military leaders view cohesion as necessarily tied to heterosexuality and homosociality.

Because unit cohesion is intimately tied to homosociality, heterosexuality has been perceived as a necessary condition. Homosexuality disrupts the military order because it creates tensions and puts in peril the homosocial relations that are at the core of camaraderie of the troops (cf. Osburn and Benecke 1997; Steans 1997, 93). Furthermore, homosexuality questions the definition of heterosexual masculinity, the entrenched norm of an institution of hegemonic masculinity. Of relevance here is also the homophobic way 'manly women' have been viewed in the military, as discussed earlier in this chapter. Other indicators are the way homosexual officers and conscripts have been treated over the years.

The aversion against homosexuality remains explicit in many militaries, including the Turkish (Biricik 2011) and, as pointed out earlier, the U.S. Armed Forces. Discrimination on the grounds of sexuality has been illegal in Sweden since 1999, and homosexuality has been officially accepted since 1979; prior to that it was considered an ailment. Despite this, homosexuality was for a long time a completely silenced topic and remains a problematic issue in the military. The secrecy and silence around homosexuality in the Swedish context is remarkable and goes far back, according to Sarah Mared (2008), who researched the Swedish military archives for dismissals or other narratives on homosexuality. She concludes that homosexuality seems to have been met with silence officially and with homophobia in practice.

A debate on sexuality began in the SAF in 2000 when Krister Fahlstedt (2000), an officer, published the first study of homosexuality in the SAF. It coincided with the instatement of the Homosexual Ombudsman (HomO) and the emergence of a lobby organization for homo-, bi-, and transsexual military staff (HOF Board 2010). The same year, High Commander Johan

Hegerstedt publicly condemned and denounced any type of discrimination against sexual preference in various fora. Since, SAF participates in the yearly Pride festival in Stockholm. Yet, changes have been slow to come. During 2003 the government commissioned the first comprehensive survey on the working conditions of homo-and bisexuals. One result of the survey was that male-dominated workplaces, such as the military and the policy force, were less tolerant toward homosexuals than other work sectors (Bildt 2004). Tolerance of sexual difference is now part of the strategic goals of the military, for example in the 'new values for the defense'.

POSTNATIONAL MILITARY STRATEGIES AND THE VALUE OF DIFFERENCE

In a parliamentary debate in December 2004, Former Defense Minister Leni Björklund, also the first female defense minister in Sweden, was asked about the importance of popular support for the new defense strategies. She argued that national identification and democratic accountability, previously considered crucial, would also be important in the postnational defense. The vision for the new defense was, she argued, that it "must reflect contemporary society." She continued by saying that also the military organization "must reflect our multicultural society that has as its objective the equality between men and women." The defense minister's vision was of the military as an integrated part of civil society with societal values central to its activities. To her, this meant "that the Armed Forces must increase the number of women and immigrants in the ranks and welcome homo- and bisexuals." (*Kammarens protokoll* 2004, my translation). Clearly, there is an ambition to reform the military organization toward a postnational defense and to deal with difference as a part of this agenda. This section investigates the strategies that the military has launched over the years to deal with gender and sexual difference, with an emphasis on the most recent, and asks: How does the military construct gender and sexuality in strategies for the postnational defense?

The vision of the Defense Minister was endorsed in the governmental bill of the same year (Proposition 2004/05:5, para. 7.5.14). Defense Minister Björklund articulated what later became part of what is called 'new values for the defense' (*Försvarets nya värdegrunder*). High Commander Håkan Syrén took Björklund's ideas to heart and presented the visions and strategies for the future military organization in a report titled "The Way Forward" (*Vägen framåt*; SAF 2005). Syrén also announced that the SAF was now beginning an extensive effort to change behavior and attitudes to eliminate discrimination through information campaigns, education, and networking. During 2005 Syrén visited all the military units in Sweden and introduced this new policy (SAF 2006).

There are many reasons to be wary about the prospect of change in military institutions. Organizational and institutional theory have often pointed to problems of institutional inertia and embedded norms affecting prospects and attempts at institutional change (March and Olsen 1989). In an interview, Krister Fahlstedt argued that the ambitions set out in the military's values and strategies are not easy to carry out in practice because of the strong prejudices against homo-, bi-, or transsexual persons within the Swedish military. The conscript unions' many reports from the field confirm this: Homophobia continues to be part of daily jargon (Värnpliktsrådet 2008b) and sexual harassment prevails (Värnpliktsrådet 2009). As evidence of resistance within the military organization toward educational efforts related to the 'new values', a practice has developed whereby one can declare a time-out. It has evolved into a hand signal: with one hand you make the V and with the other hand the lid or the T for time out (Pettersson 2008, Värnpliktsrådet 2008b). This is a form of resistance to the new ideas that are being taught in the military training of the postnational defense. The educational efforts are perceived as imposed and politically correct. During 'new values' time-out, sometimes even initiated by officers, recruits are able to relax and go back to 'normal' behavior. As has been argued in this study, 'normal' in military institutions refers to the norms of hegemonic masculinity.

The strategies that the military organization proposes to deal with the problem of difference are interesting because they are an example of the military's attempt to transform itself toward a postnational defense agenda. In this process it also becomes intelligible how the military views difference and how it conceptualizes women and LGBT persons—why they should be part of the military and what they can contribute. Three different arguments for their inclusion emerge. First, the "representation argument" assumes that the military should reflect the population that it must defend. Second, the "respect for the rule of law argument" means that legislation and regulation relevant for a democratic society should be applied also in the military. Third, the "diversity argument" is concerned with the resources that are lost if not all groups are included. As demonstrated in the following, these arguments have been applied, not always consistently, to argue for the inclusion and acceptance of women and LGBT persons in the military.

The representation argument is as follows: If the military is to reflect a diverse society, it has to be representative; that is, whatever share of people the group has in the population should be reflected in the military. In a democratic society, it may be the only way for a defense to acquire the broad support it needs to carry out its activities, and it reflects the vision of the former defense minister. However, the idea of being a representative organization rests a bit uncomfortably with the military, particularly because it was previously charged with the task of maintaining an efficient and effective organization geared to the demands of war. Its interest has been in acquiring the competence required to respond to this task. At the same time, the military recognizes that to appear as an attractive employer, it must be in tune with the values in society. This is likely to be even more relevant for the Swedish military now that conscription has been suspended.

Gender has been discussed only in the language of representation via voices outside the military, particularly those who have argued for gender-neutral conscription. The representation argument is common in the public debate in Sweden regarding gender, exemplified by the call for equal representation in political parties and executive boards of companies. A representation of 50 percent women and 50 percent men is seen as the ideal and 40/60 as acceptable. The military has not discussed such ambitious goals but has hoped to include a few women. In the equality strategies, there are no set target quotas for women (SAF Equality Plans 2004, 2006, 2009). Informal goals to increase the number of women have been set occasionally, for example, 20 percent in 2008 for the Nordic Battle Group and 15 percent among 2010 recruits in the Blekinge Wing (Värnpliktsrådet 2008a), but these targets are far from an equal representation. When it comes to sexuality issues, there has been no argument that the military staff should reflect the percentage of LGBT persons in society.

The respect for the rule of law argument is one that has been particularly emphasized in the 'new values for the defense'. Societal laws against harassment and discrimination and for the respect for human rights must be adhered to also in the military organization. The military tends to be viewed as more equal to other state institutions when the extraordinary conditions of war or the threat of war are less frequent. To the individual soldiers, the freedom from discrimination and the tolerance of differences are important. The conscripts' union's major concern since the early 1970s when it was founded has been to protect the rights of conscripts against punishment, abuse of power, and harassment. Sexuality, especially LGBT rights, are perceived strictly in terms of the respect for the rule of law. Sexual preference is viewed as a basic human right within the military organization. Thus, to discriminate on the grounds of sexual preferences among consenting adults in the workforce or in society is illegal. Sexual harassment carried out explicitly and on an interpersonal level against women has been articulated as a breach of the rule of law, and systems to deal with complaints have been set up. Harassment toward women due to their sex (e.g., withholding information, ignoring) is not in the same way regulated and monitored, nor has it been discussed as much as sexual harassment in light of the rule of law.

The final argument is about diversity as a resource. In general terms, the military has often talked about the added value of diversity among its staff. In relation to women and gender, this argument has a long history. The military often equates gender with women and has considered women in terms of the function they have in the organization, asking, for example, whether women contribute with skills otherwise lacking or whether they can make up for the general lack of manpower. I have found no evidence of a similar argument that LGBT persons would be a resource that could be put to use for better military performance. When the 'new values of the defense' were first articulated in 2005, ethnic diversity was argued to be, like gender, a resource, something useful to the SAF because it can lead to higher quality and a readiness to solve problems within a military organization increasingly involved in international missions in distant locations. This argument about ethnicity as diversity has not been followed up.

The resource argument often surfaces with discussions on women's inclusion in the military organization. Political documents of a general nature on women's role in the military typically stress that women and men are similar and, hence, argue that men and women should have equal access to the military on democratic grounds. However, in the military organization women have often been perceived in terms of possessing different characteristics and skills. Owe Wiktorin, a former chief commander, was quite clear about this when he spoke to a group of female officers and said that women's qualitative contribution to the SAF "requires that you female officers fulfill your role as officers while retaining your female identity and with the norms and values you have as women" (Defense Ministry 1995, 49). The complementary knowledge women provide and their positive impact on the atmosphere of the SAF has been stressed. Examples are statements such as "Women contribute with a *different* way of thinking, a *different* way of talking, and establish other kinds of relationships within the defensive forces. Women can develop better communication within the defense"(*Värnpliktsnytt* 1995, 11, my translation, italics added).

A program initiated by the SAF in 1990 to resolve problems with sexist attitudes and harassment looked at difference as a source for creativity, new ideas, and new ways of working. The program, called Kreol (an abbreviation for *kreativ skillnad* [creative difference]) had as its base beliefs that difference between men and women is biological and that women and men's brains are different, which explains why they have different communicating styles as well as different ways of leadership. The program was developed by the management and consulting sector. This difference was perceived as 'good' in that it provided a needed resource (Göbel 2000; SAF Internal Document 1995, 1996). The diversity argument continued to be highly relevant even after the program ended in 1999. For example, it was used to coax reluctant men into favoring the recruitment of women to the SAF: Women would be contributing with something different that the military organization needed but was lacking. This view lingers even though the official position has changed and the Kreol program is no longer in effect (Ivarsson 2002; Weibull 2001). Through personal correspondence, I found that gender training in the SAF in 2009 had as required reading *Egalias Döttrar* [Egalia's daughters], a book by Gerd Brantenberg published in 1977. Although the book is feminist and criticizes patriarchy, it does so by completely reversing and humoring the gender order, turning patriarchy into a matriarchy, hence reinforcing the difference argument. In relation to the role of women in peacekeeping, difference has been a frequently used argument for why women are needed in international missions. Female peacekeepers are presumed to possess certain skills because they are women, and these types of skills are believed to be needed and a resource in peacekeeping efforts.

The diversity argument is problematic because of the way it accentuates difference. This argument, although it may at times be a useful strategy, tends to fix certain characteristics, behaviors, and competences to those who are expected to carry the diversity. It assumes that while there is certainly difference in society, some individuals are more different than others. The rationale for including these 'different' individuals and groups in the organization hinges on the fact that they will actually behave, act, and provide the diversity they are expected to embody. Through the diversity argument, the majority is simultaneously constructed as homogenous and as naturally associated with the organization. The way that the Swedish military organization has conceptualized difference, mainly as respect for the rule of law and as an added resource, suggests a certain rigidity or inertia in this institution of hegemonic masculinity, reluctant to give up the notion that heterosexual men are the 'real' protectors. The fact that almost all of the initiatives that have been launched by the military to deal with new values and norms, harassment, gender equity, and creative difference have been conducted in project form, rather than as part of the military's regular activities, further accentuates this point.

CONCLUDING COMMENTS

This chapter has explored gender and sexuality in the context of the Swedish military and defense organization. The goal was to understand how those 'differences' play out against deeply embedded norms of hegemonic masculinity and to discover how gender and sexuality is part of the construction of such institutions. The chapter began by taking the perspective of the woman soldier involved with the practices of the military. It made the masculine norms apparent while at the same time showing the dilemmas around the formation of a woman-in-arms femininity. The analysis then explored the way sexuality is used as a resource in the military by a focus on unit cohesion, believed to be essential for military performance. The finding was that unit cohesion has been constructed on two exclusionary mechanisms of male heterosexuality and homosociality. These are threatened by the inclusion of women and homosexuals and can explain why there are military policies that exclude women from combat and homosexuals from serving. Finally, the chapter explored attempts by the Swedish military organization to deal with sexuality and gender as embraced in the 'new values for the defense'. These values are in line with what would be expected of a postnational military, discussed in the next chapter. It is interesting to note that the 'differences' discussed—sexuality and gender—were dealt with according to different logics or arguments. Sexuality was considered in relation to the respect for the rule of law while gender was considered a resource. This logic appears less challenging to the military as an institution of hegemonic masculinity. To treat women's presence in the military as a matter of providing complementary skills that are otherwise lacking assumes an essentialist understanding of gender (i.e., that men and women are different and have different functions to fill and contributions to make to the military). Furthermore, it suggests that women are welcome or interesting to the military organization only if they actually contribute the skills they are assumed to have as women. This is, as we will see in a later chapter, a view that has been confirmed in international operations but remains highly problematic.

CHAPTER 3

The Postnational Defense and the Cosmopolitan Military

A postnational defense focuses on the security situation outside the nation's borders, often in a multilateral context. An example is the Swedish postnational military that appears to be guided by a cosmopolitan vision, which has resulted in an emphasis on peacekeeping activities and conflict prevention. This chapter poses questions regarding the relations between such a postnational defense and the military organization. As scholars have argued, there is a new relationship between security politics and the military (Coker 2008; Viktorin 2005). Elliott and Cheeseman's (2004) theory on cosmopolitan militaries is helpful for understanding this analytically. Its starting point is in cosmopolitan ethics, and it is directly related to military organization and practice. Moreover, cosmopolitan ideals are useful because they suggest that the military can be denationalized and democratized.

Cosmopolitan values may even have the potential to demilitarize the military, or as Annika Björkdahl (2005, 222) states, "There is room for military means in promoting a cosmopolitan vision, but for defensive, protective purposes such as prevention and not as traditional war." This is highly compelling to a feminist analysis and responds in some way to the appeals of Laura Sjoberg (2006, 205–213) to formulate feminist ethics of war. Having said this, regarding the conditions for change and the background of the national military organization, the transformation is not expected to be smooth or easy. This chapter is formulated around three challenges. The first challenge is to denationalize a military that has been tied to the politics of the nation, its territory, and its people and to focus on the well-being of people in faraway places and show solidarity with them. The second challenge is to demilitarize, or take a military organization that has been geared toward war and combat exclusively throughout history, and turn it into an organization that mainly 'does peace'. Combat certainly seems far from practices like peacekeeping and conflict prevention. A third challenge is to democratize the military, an institution of hegemonic masculinity closely tied to the power of men in a strict hierarchical fashion. The previous chapter outlined the extension of that challenge. This chapter will discuss these challenges in relation to the Swedish postnational defense and to the experience of militaries of other midsized powers that are also embracing cosmopolitan-like values.

THE COSMOPOLITAN-MINDED MILITARY

Cosmopolitan militaries have been labeled "forces for good" (Elliott and Cheeseman 2004) because they adhere to cosmopolitan norms. *Cosmopolitanism* is a normative political theory that focuses particularly on the common values and bonds among humans regardless of borders and territories. The respect for human rights is central.[1] Ideally, a key ambition of a cosmopolitan is to protect human beings under threat elsewhere, outside national boundaries, and to save 'distant others', in the name of human rights. Cosmopolitanism is postnational due to the respect for humans across national boundaries and the emphasis on their common destiny (Beck 2006; Held 1995). The task for a cosmopolitan military is to defend 'the other' rather than to defend against 'the other', which is the objective of the territorial defense (Elliott and Cheeseman 2004, 2–3). A distinction between inside and outside the state cannot justifiably be maintained. Humans are bound together as a community, and the cosmopolitan military cannot be content with protecting nationals—that is why it is involved in peacekeeping and conflict prevention. It engages in different forms of international peace enforcement and peace keeping (cf. Smith 2007; Fine 2006). For Sweden, to develop a cosmopolitan military and a postnational security strategy does not represent a complete break with the past. Rather, it builds on the neutrality strategy. Yet many aspects of the military in the postnational defense must be considered new trends.

The emphasis on international peacekeeping and the changes in civil–military relations as well as the role of the soldier are examples of a transformation that Elliot and Cheeseman (2004) and Kaldor (2001) have observed in the defense sector of many industrialized countries and labeled in terms of a cosmopolitan-minded defense and military. Such militaries are "deployed for cosmopolitan purposes" and would "be expected to have a broader span of rules, functions and responsibilities than do conventional defense forces" (Elliott and Cheeseman 2004, 278). In turn, "these broader, other-regarding and security-building activities are likely to require . . . that the soldiers deployed possess skills and attributes that extend well beyond the values and duties normally associated with the profession of arms." (Elliott and Cheeseman 2004, 278; see also Joenniemi 2006b, 10). The assumptions are based on the observation of a general diversification of military practice among state militaries. Few state militaries of industrialized countries are engaged in defending their territory at home—a large number are active in distant places with a focus on peace enforcement, peacekeeping, and crises management activities. This is the case in Sweden.

FROM NEUTRALITY TO SOLIDARITY WITH DISTANT OTHERS

As a small state, Sweden had little choice but to adjust to the international balance of power during the Cold War and act as a buffer between the two superpowers. The strategy was to be nonaligned during peace time and remain neutral in case of war. Compared to the cost of war, Sweden's neutrality choice made good economic sense. As argued in chapter 1, neutrality was not a pacifist strategy nor was it free from militarism or nationalism. It was a defensive strategy that required investments, economic resources, and a substantial defense organization (Logue 1989, 55–58; Syrén 2004, 10). Neutrality was supported by a domestic weapons industry in order that Sweden would not depend on foreign military powers for its military and defense needs, and considerable investments were made in both research and the production of arms. Furthermore, neutrality was understood as requiring a high degree of self-sufficiency with a highly subsidized textile and agricultural sector and thus became part of many productive sectors of society (Dohlman 1989).

The new international security situation that began at the end of the Cold War presented an opportunity to develop a postnational defense and reevaluate neutrality. In April 2001, late Foreign Minister Anna Lindh announced that Sweden would no longer call itself a neutral state, and in the declaration of the government in February 2002, the new security approach was confirmed (Statement on Government Policy 2002). Although studies have revealed that neutrality was more robust in theory than in practice, it was resilient and an integral part of Swedish national identity (Agrell 2000, 140–181; 1991; Holmström 2011; SOU 1994:11). As we have seen, neutrality constructed a national identity (Goetschel 1999, 116; Kite 2006, 107), tying security and defense politics to *folkhemmet*. Neutrality was a way to create a sense of national pride and grandeur (Eduards 2007,

37), and because it touched on sentiments related to the nation, it became credible to the citizens (Sundelius 1989, 1). Neutrality, to be a viable security strategy, had to consider others' perceptions of Sweden (and particularly other state leaders; Åström 1989).[2] These leaders had to be convinced that Sweden was serious about nonalignment and would stay neutral in the case of war in the region. Hence neutrality was part of Sweden's projection of a specific image of itself into the world. It provided an in-between position, a particular national identity that also seemed to make Sweden a 'winner' internationally (Agrell 2000, 278; see also Ingebritsen 2006).

There were seeds of cosmopolitanism already in the Swedish neutrality policy. While it was clearly national in stressing the need for a strong territorial defense, mass mobilization across the population, and not physically intervening in other states' conflicts or taking sides, there were also cosmopolitan-like elements. Sweden was involved in development cooperation and in international missions and pursuing what Elgström (1982) calls an active foreign policy, often representing the interests of smaller and nonaligned states in international relations, acting as a mediator, and strongly supporting international cooperation through the United Nations (UN).[3] Sweden's legitimacy in this context was facilitated by its status as a neutral state, the absence of a direct colonial history, and its relations with developing countries. This image of Sweden's role in the world continues to be significant, especially when postnational defense efforts are being concentrated outside the borders in the international scene (Bailes 2006, 16–22). Maud Eduards (2007, 56–57) suggests that as the contemporary Swedish military is conducting its peacekeeping tasks, it continues to project this image of Sweden in the world. Actually, the military organization puts pride in being the 'world's' conscience' and particularly in being the world's most gender-equal country.[4] The Swedish military and its peacekeepers lead the way in representing this image of the gender-progressive nation (Eduards 2011; Ivarsson and Edmark 2007, 19). The Swedish male soldier carries forth the 'new' masculinity; it portrays him as equal, caring for his children and the home and supporting gender politics.

What are the grounds for calling the Swedish defense denationalized? Annika Bergman (2006) suggests that Sweden, like other Scandinavian countries, continues to have a clear and consistent internationalist component in its foreign and security policies (see also Miles 2006, 77) and that this component has continued to guide the way Sweden views itself in relation to other states (Bergman 2006, 76). Sweden's internationalism is based on a social democratic ideological component common to Scandinavian foreign policies (Bergman 2004, 2007; Patomäki 2000, 116). It is the outcome of "democratic, solidaristic and egalitarian values and interests, that albeit social democratic in origin, seems to have been deeply institutionalized" (Patomäki 2000, 132). It has become an element of what Åselius (2005) calls Sweden's strategic culture. Thus, the postnational security and defense represents a continuation, not a clear break from, security strategies developed in the past and exemplified by the emphasis on peacekeeping, conflict prevention, and peace enforcement in the security strategy ("Statement of Government Policy in the Parliamentary Debate on Foreign Affairs 2006, 2010). It resonates with Sweden's security policy's objectives as articulated by the Swedish Defense Commission, which is responsible for overall security strategy. The objectives are to assure the country's peace and independence, to contribute to security in the region, and to strengthen international peace and security (Ds 2003:8; Ds 2004:30; Ds 2007:46). Sweden, like other Scandinavian countries, has taken the internationalist agenda further by acting as an example of an entrepreneur by "collectively strengthen[ing] norms of conflict mediation and peaceful resolution of conflict" (Ingebritsen 2002, 17; see also Björkdahl 2002). In this respect, Sweden is not unique.

OTHER COSMOPOLITAN-MINDED MIDSIZE POWERS

Sandra Whitworth (2004), who studied Canada's peacekeeping forces, argues that midsize powers like Canada, Scandinavia, and the Netherlands participate in international missions because this is the field available to them. It is a way for them to legitimize their militaries, while at the same time it legitimizes these countries as nations. The internationalist dimension seems to increase in relevance as Sweden takes part in peacekeeping missions, conflict prevention, diplomacy, and mediation efforts both in the European Union (EU) and the UN. For the EU, the principle of solidarity in the EU Lisbon Treaty, in effect since December 2009, certainly emphasizes the direction toward multilateralism as it calls for solidarity between the member states to support each other if attacked by any outside aggressor. For some countries, peacekeeping has become the *raison d´être* and a way to reconceptualize the projection of a national identity. Sandra Whitworth (2004) says about Canada that "peacekeeping serves as one of the 'core myths' of Canada's 'imagined community'." It is a myth that "locates Canada as a selfless middle-power, acting with a kind of moral purity not normally exhibited by contemporary states" (Whitworth 2004, 14). This resonates well with the Swedish projection of itself in the world and also with other Nordic states. Participation in international missions "forms a natural part of the Nordic states' self-images" (Stamnes 2007, 453). What we may be seeing is that through a postnational ambition another, different national identity is being formed in which cosmopolitan values serve as core ideas.

Interestingly, the Finnish military organization, which Pirjo Jukarainen (2011, 136) calls a "peacekeeping superpower," has also turned its cosmopolitan ambition into a way to improve the capacity of the military. To go abroad on peacekeeping missions serves as a "training camp," a way for the military organization to obtain the necessary training for the troops. As articulated in a governmental report (Skrivelse 2007/08:51), the Swedish military's tasks are twofold: to engage in cosmopolitan-like activities but also use the experiences from such missions to increase the military's capabilities to defend the territory. Similar to Finland, international missions thus also serve as training to prepare the military.

The role of the Dutch military in the context of the nation can be compared to Sweden's. Stefan Dudink (2002, 149) writes that "the country's lack of military credentials is presented as a virtue rather than a weakness." It means that the Netherlands sees itself as a moral nation, highly qualified in areas such as peacekeeping. However, in contrast to Sweden, the Dutch peacekeeping troops were subject to the trauma of Srebrenica, when around 8,000 Muslim men were killed basically in front of their eyes (Zarkov 2002). The mission allowed them to use weapons only to protect themselves. It severely questioned the capability of the Dutch military and undermined their legitimacy, labeling them as "too sweet and innocent for war" (Sion 2006).

The Dutch Srebrenica 'trauma' surfaced again in 2010 in the context of a U.S. Senate hearing on lesbian, gay, bisexual, and transgender (LGBT) persons in the military. During the hearing, a retired U.S. general and former North Atlantic Treaty Organization (NATO) commander made a statement against including homosexuals in the military. He based his position on the Dutch's experience in Srebrenica and argued that European militaries have changed by becoming democratized and demilitarized with less focus on combat. In other words, as militaries allow homosexuals, as most of them do, they lose their potency. Because the Dutch military had in this way been "weakened," the troops could not perform, and this was the reason for the massacre in the U.S. general's view. The general's view was forcefully criticized, as the Dutch military and political establishment protested the accusation and called it nonsense ("Dutch Fury" 2010; "Gay Dutch Soldiers" 2010; Naughton 2010). The U.S. general, nevertheless, gave voice to opinions that still have standing, at least in parts of the military establishment, that the decline in combat readiness and training leads to a decline in military performance.[5] Combat is indirectly coupled to a militarized masculinity, that is, threatened by the presence of homosexuals, which suggests that Dutch and European militaries are 'sissies' as compared to the U.S. military.

TASKS BEYOND MILITARY-CIVIL BOUNDARIES

The relative increase in the importance of international activities compared to other duties of the military is notable in the Swedish context. Bergman (2004, 181) concluded from her analysis of the Nordic militaries that they have many cosmopolitan features and that their "participation in multilateral security and defense cooperation supports this contention." The Swedish military has increased its emphasis on peacekeeping operations. For example, in 1986 only 2 percent of the soldiers were designated for international peacekeeping operations (Dörfer 1997, 43), and it was "seen as a sideshow by the Swedish military" (Wedin 2006, 142). The other Nordic countries had a similar perception of the importance of peacekeeping to the military, yet about 25 percent of the peacekeepers have come from the Nordic countries (Jakobsen 2007, 458) and the importance attributed to international missions had increased in the Nordic countries as well. In 2005, 7.5 percent were designated for EU operations alone. Apart from serving in the International Security Assistance Force (ISAF) and UN peacekeeping forces, Sweden contributed with the Nordic Battlegroup in 2008 and again in 2011. Since 2010, all soldiers must accept to go on international missions. The military organization's aim is to try to develop deployment units that are flexible so that staff and equipment can be used in both the battle groups and in peacekeeping operations in a range of different tasks and situations.

A cosmopolitan agenda puts demands on military activities and training. Elliott and Cheeseman (2004, 4) argue that militaries "deployed for what are essentially cosmopolitan purposes are expected to perform a range of tasks, some of which may fall outside their traditional ambit." This trend began in Sweden in 1996 (SOU 2001:23, 46), and, since then, the military organization has been organized in a way that reflects what may be expected of a cosmopolitan military. The Defense Resolution of 2004 further outlined this (Proposition 2004/05:5). The Swedish Armed Forces (SAF) is to focus on an active, operational defense. In the event of a territorial invasion, defense will occur through flexible and rapid deployment units (cf. Perspektivstudien 2009). The same forces are expected to be deployed mainly outside the borders in peace-promoting missions to defuse international conflicts. Hence, they need to be ready for combat as well as for peacekeeping and peace building. This is in line with what Elliott (2004, 26) argues: A cosmopolitan military must be continuously prepared for a whole range of eventualities. It must be rapidly deployable and trained for both combat roles and peacemaking roles. This can be seen in the Swedish ISAF operations in Afghanistan but also in other peacekeeping operations, for example, in Liberia and Kosovo (*Insats & Försvar* 2006, no. 2). The ISAF is to oversee demilitarization by searching for mines and loose ammunition as well as maintaining security. Tasks involve escorting the national police, assuring the safe arrival of development aid, and securing the free passing of materials through the area. At times it involves aiding and training the local police forces or facilitating communication between different local actors and groups (*Insats & Försvar* 2006, no. 4). While the primary task of the ISAF in the provincial reconstruction team is to assist local leaders in building a secure and safe society through security sector reform and encouraging good governance (Olsson and Tejpar 2009, 30), there are many risks and threats. Thus, they must be prepared to use military means if needed.

Being cosmopolitan minded, the Swedish military has adapted to demands for activities that contribute to peace and security abroad. This does not mean that the resources previously devoted to national and territorial defense have been redirected to peacekeeping tasks. The military organization has also been considerably reduced through budgetary cuts. The fact that Sweden now puts fewer resources into the military sector can in itself be interpreted as demilitarization, because the resources then become available for use in other sectors of the welfare state. At the same time, it might actually speak against the Swedish military as cosmopolitan because it is not taking its security ambition 'to save distant others' seriously (cf. Wedin 2006, 146–148). Linda Åkerström (2008), who writes for the Swedish Peace and Arbitration Society, is critical of the way military spending has been communicated to the public. She finds it paradoxical that Sweden is supposed to have a severely reduced defense while the costs remain high. The explanation is, she argues, that old military structures remain in place, and the close relationship with the weapons industry has led to inefficiency when it comes to the acquisition of military equipment (Åkerström 2008, 28–31).

While cosmopolitan military ideals may be more desirable from a post-national and a feminist perspective, they remain nevertheless, paradoxical. The deployment of a cosmopolitan force, argues Elliott (2004, 24), goes to the heart of and challenges the military rationale, the military's mission, and the rules applied in military engagement. We may ask: does it go far enough? Indeed, today when armies are sent into battle they are increasingly dispatched as peace enforcers or peacekeepers. This trend is ironic in relation to the classic dualism of war and peace and is a sharp break with the notion that the military is about war. The rationale of state militaries has been tied to their monopoly on the use of organized violence. Recall again that militarism, or the values that have guided military activities, have considered armed violence and the threat of it as a way to solve problems related to the security of the state (cf. Jabri 1996). The same organization is now in charge of enforcing or assuring peace.

That cosmopolitan militaries should be prepared to use a wide range of means, from military methods to policing, collaboration, and communicative acts that can be seen as overstretching the meaning of cosmopolitanism, becoming more closely associated with Kantian peace ethics. Dexter and Gilmore (2006, 9) say that cosmopolitan militaries "require a shift away from the traditional military practice of warfighting towards methods and tactics better suited to the protection of human security." This does not automatically exclude the use of military violence. Hence, the question of the use of violence in cosmopolitan militaries is a pertinent one. The focus is on human security and human rights, and this requires constant political judgment in the face of violence. There is this deep sense of ambivalence on the one hand that military intervention and the use of military force is a responsibility when severe crimes against humanity are committed and, on the other, of the unintended consequences of the use of violence and military power in such circumstances (Fine 2006). For cosmopolitan militaries, it is not that violence is completely inappropriate; the question is when should violence be used, on what grounds, how extensively, and in what way (Buchanan and Keohane 2004; W. Smith 2007).

Making these kinds of judgments is also what differentiates among militaries and determines whether they are cosmopolitan minded. Defense Minister Tolgfors (2009) argues in relation to this in a debate around the accelerating violence in Afghanistan. He underlines how different the Swedish peacekeepers are in comparison to those of the U.S. or German armed forces. They are different in their view on how and when violence should be used as well as in their attitudes toward civilians. Perhaps they are also different from their Scandinavian neighbors. The Danish prize-winning documentary *Armadillo* (2010), directed by Janus Metz Petersen, portrays how the Danish peacekeepers stationed at Camp Armadillo changed. When faced with violence and local conditions, they failed to stick to cosmopolitan norms, were disillusioned, and turned violent. The documentary certainly calls attention to the question of whether it is possible to make peace via military means. The situation in camp Armadillo in the green zone is very pressing ("De vinkar inte" 2009; Jensen

2009; "Varför är vi i Afghanistan?" 2009) and indeed tests the limits of the cosmopolitan-aspiring military.

Another paradox is that in a time when Sweden's own security situation is more relaxed and free of military threats, the risk that the Swedish military and, more particularly, its trained soldiers and officers will become engaged in military activities has increased (cf. Bolin 2008). During recent years, there have been debates over Sweden's involvement abroad in peace-keeping operations; the Swedish presence in the ISAF in Afghanistan has been particularly controversial ("Allt farligare i Afghanistan" 2005; "Ännu en svensk Afghanistansoldat död" 2005; "Förhöjt hot mot styrkan i Afghanistan" 2005). This is due to the resources used, the danger of the situation for military staff, and the resulting precarious strategic/military situation. Peace efforts by the ISAF are carried out alongside the war con-ducted by U.S. forces against the Taliban in southern Afghanistan. The involvement of Swedish peacekeepers in the ISAF in Afghanistan has proven to be particularly risky. Two Swedish peacekeepers were killed and another two injured by a bomb while on duty in Mazar-i-Sharif in Decem-ber 2005. The Swedish ISAF have had to defend themselves in open fire against criminal groups and the Taliban ("Brottslingar attackerar svensk styrka" 2006; "Svenska soldater beskjutna i Afghanistan" 2006; "Svenska soldater i eldstrid i Afghanistan" 2006; "Svenskar Under Attack" 2006). During 2008 and 2009, more peacekeepers were sent to the Camp North-ern Lights, and security measures around the ISAF base were increased sig-nificantly (SAF 2009). In 2010 another three Swedish peacekeepers based at the camp were killed. As the violence in the area increases, so do the security threats to the ISAF and the local population ("Sorg i lägret efter svenskens död" 2010).

Defense Minister Sven Tolgfors said the work performed by the Swedish ISAF in Afghanistan showed "international solidarity in practice" (Tolgfors 2009), and it is generally understood that the military must be prepared for higher risks in the future. Thus, contrary to what might be the general under-standing, peacekeeping is neither less risky nor does it necessarily require fewer resources to be invested in the military. The risk to the individual sol-dier and the unit may increase as the military becomes more engaged in the international setting. It can be argued that this is to be expected of a cosmo-politan military with a goal of 'saving strangers.' For the Swedish ISAF forces in Mazar-i-Sharif, it entails more risks and more violence.

However, it is unclear what type of sacrifice can be expected of the peacekeeper to effectuate a solidarity with 'distant others.' For cosmopoli-tan militaries, the central idea is that we are joined together as humanity and that human life is equally valuable. The expectation of the Swedish soldiers today is that they be prepared to sacrifice for 'distant others' and accept that the tasks they perform may not be strictly military but more like civilian duties. As Elliott (2004, 28) suggests, "Military personnel engaged in cosmopolitan missions are expected to risk their lives not just, or even, for their co-nationals but for humanity as a whole." Mary Kaldor (2001, 131) explains, "Whereas the soldier, as the legitimate bearer of arms, had to be prepared to die for his country, the international soldier risks his or her life for humanity". This shift means that cosmopolitan soldiers should be ready to sacrifice themselves for unknown strangers. Following the death of the first two Swedish soldiers in Afghanistan in 2005, former Defense Minister Leni Björklund stressed that Sweden will have to accept and be prepared for these kinds of losses in the contemporary security context (Speech Sälen January 2006). Judith Butler (2004, 2009) argues to the contrary about war, saying that in war it is clear that life does not have a universal value. Life is valued and differentiated in terms of national belonging, ethnicity, class, and gender status. It may be that when the stakes become higher, cosmopolitan values seem less relevant.

The cosmopolitan military as it has been discussed here suggests a demilitarization in that the tasks soldiers are expected to master are wide ranging, from combat to 'talking with locals.' Matching the appropriate means to the situation is a delicate task, particularly when it is about deciding whether to use violence for the purpose of peace. Yet violence seems necessary in some situations, and this is also a peril for the individual peacekeeper who is asked to risk and sacrifice his or her life for the good of humanity.

MASCULINITY AND THE COSMOPOLITAN MILITARY

Cynthia Enloe (2007, 40) writes that "the more militarized the understanding of what national security is (and what it is not), the more likely it will be that the conversation about national security—and international security—will be a largely masculinized affair." Does the contrary hold? If a cosmopolitan military is less militarized and less nationalized, does it imply that it is also less masculinized? Christopher Coker (2001) suggests that would be the case. Military scholar Van Creveld (2001, 2008, 395–409) also speaks to this and argues that Western militaries' increasing participation in intervention for peace enforcement and peacekeeping is evidence of a feminization.

Sandra Whitworth (2004) studied the transformation of the Canadian military. The Canadian military relies on soldiers who have been trained to fight wars to conduct peacekeeping tasks. She describes it as follows: "The blue-bereted peacekeeper is supposed to be benign, altruistic, neutral and capable of conflict resolution in any cultural setting—a warrior prince of peace" (Whitworth 2004, 12). She sees a problem in that traditional military education and training focus on preparing for possible combat situations rather than peacekeeping activities. The military organization trained Swedish conscripts during neutrality to defend the Swedish territory with military means against an attack in situations of war. Recalling that conscription training constructed ideas about the citizen as the white heterosexual man, the recruited soldier has traditionally been encouraged to develop aggression and strength and to rid himself of anything that resembles a female attribute. In peacekeeping, this gets turned on its head: Violence is failure, and the path to peace is cooperation and reconciliation (DeGroot 2001, 34). Hence, the military relies on soldiers who have been trained to fight wars while peacekeeping requires that the soldiers reject many of the traits they have been trained for as soldiers. Peacekeeping seems to demand a new type of soldier, one that can be prepared to do both war—in peace enforcement and security tasks—and peace (Fox 2001, 20). Sandra Whitworth (2004, 151) is doubtful whether the military, as a strictly ordered hierarchical organization and whose main purpose is "the creation of men . . . who will be warriors, who are prepared to kill, and die," is appropriate for carrying out "missions dubbed peacekeeping." In other words, are they up to performing the 'civilian-like' tasks they are asked to do?

Recalling that gender is performative and always something in the making, militaries have been busy 'making boys out of men' (i.e., 'doing masculinity'). The construction of masculinity has been intricately tied to military practices. Higate and Hopton (2005, 443) write: "The reciprocal relationship between militarism and masculinity functions at the level of identity as well as the state." Thus military training does its part in constructing masculinity through combat exercises and in relation to carrying weapons, learning to shoot, performing surveillance for suspected movements, securing areas, and so on. In this training, soldiers are prepared to meet the challenges they may encounter as they are sent out on missions. Masculinity is simultaneously reinforced. The move to more civilian-like tasks, such as policing, cooperating with locals, communicating, and assisting, signify an inclusion of tasks that have been coded as feminine through military practices in the past. Hence, it may be considered a demasculinization.

The paradox is evident in the Canadian military. The image of Canada as peacekeeper represents a "benign and altruistic image"—an image "fundamentally at odds with the roles soldiers" are expecting to perform and "indeed were created to perform" (Whitworth 2004, 87). This is seen in other contexts as well. According to the research of Liora Sion (2008, 566), the training of Dutch peacekeepers "was violent and exciting," but the missions in Bosnia and Kosovo "were peaceful and more humanitarian." this made soldiers disappointed, and they concluded that "anyone can do a peace mission." Peace missions are too easy—"even women can perform peace missions, because nothing happens." In her study, Sion (2008, 561) argues that although peacekeeping offers a new military model, "it reproduces the same traditional combat-oriented mind-set of gender roles," and that is also why Dutch female soldiers "are limited in their ability to perform and contribute to peace missions." In other words, they are less fit for civilian tasks because the military has little experience with this training but also because the performance of military-type activities is tied up with the performance of masculinity.

Psychological problems such as posttraumatic stress disorder are common among soldiers who return from missions.[6] The problems may be partly a result of their perceived failings in living up to appropriate masculinity associated with militarism (Whitworth 2004, 167). Sion (2006) exemplified this when, in studying the Dutch peacekeeping forces, she found a considerable difference between infantry and artillery soldiers in the degree of satisfaction with the tasks they were to perform. The infantry or "combat units find it difficult and even undesirable to make the necessary transformation to peace missions and to abandon a combat oriented self-image" (Sion 2006, 471). While peace missions are nearly the only option for the Dutch military, infantry soldiers still envision themselves within a combat model. They have been trained and socialized with the notion that it is combat that makes the military relevant, not peacekeeping. Sion's (2006) study shows that artillery soldiers, who are used to support functions and from the beginning were not expected to be involved in combat, were also more satisfied with their duties in peacekeeping.

Christopher Coker (2007, 146) asks: "Do warriors like war too much?" It is crucial to remember that men in the military are trained to kill. Through their performance of military practice, they are trained to use violence and to 'become men'. Yet killing is also the unique skill of the profession. This control over life and death is the most important power resource of the military (Coker 2007, 61). There seem to be, at least in Canada, the Netherlands, and the Nordic countries, discrepancies between what peacekeepers are trained for—to become combat-ready men—and what is required of those deployed in international peacekeeping missions—civilian kind of tasks. If peacekeeping demands qualities like chatting and speaking with civilians, a Dutch peacekeeper said: "It is not a job for a combat soldier or even for a man—because a woman . . . could chat better" (Sion 2006, 470). In general, Dutch soldiers perceive the peacekeeping military as a "weak and feminine organization while viewing themselves as masculine and fit for combat" (Sion 2006, 455). This is what they have been trained for.

Because military performance is closely tied with the making of masculinity and the military organization is an institution of hegemonic masculinity, the failure to perform as expected because duties are perceived as feminine, or the failure to be brave, fearless, and courageous, challenges and becomes a crisis to the soldier's masculine identity. Sandra Whitworth (2008, 118) explains: "When the stoic, tough, emotionless soldier begins to feel and react, when he feels pain, fear, anxiety, guilt, shame and despair as a result of the activities in which he participated as a soldier, he violates the precepts of his military identity and can no longer fulfill the myths of militarized manhood that have shaped him." He can no longer continue his duty as a soldier.

Yet there are variations on this theme. In her study of Norwegian staff in NATO and UN military operations abroad, Torunn Laugen Haaland (2011) discusses how peace support missions conducted by a small and 'peace-loving' country like Norway relate to perceptions of masculinity. She found that there was less emphasis on the making of masculinity in comparison to Whitworth's (2004, 2008) studies. The emphasis was more on endurance, even to endure boredom, and on the need for restraint and control rather than on toughness. She also found that aggressiveness was not a commonly praised quality. Yet the forces were clear about being military men and women, and they appreciated being able to show that they could master combat-like skills. There was a kind of consensus that the base for their actions was their military skills, not their civilian-like ones.

THE DEMOCRATIC AMBITION OF COSMOPOLITAN MILITARIES

The influence of cosmopolitan values on the military organization demands qualitative changes, because cosmopolitan militaries are different from national militaries regarding their values and their practices (Elliott 2004, 18–19). Militarism values hierarchy, obedience to authority, and uniformity, which contradicts the idea of a cosmopolitan military in which democratic values are essential features. In sum and applied to militaries, the cosmopolitan imperative requires that the military organization defend and promote cosmopolitan values when it performs its functions and that the internal organization of the military supports and enforces such values. This way the military organization can set an example for good and democratic governance when deployed in places where this may be lacking. It can extend its role beyond protecting individuals from the violation of human rights to encouraging transitions to a more human system of global governance (Elliott 2004, 25). The cosmopolitan military is part of a more activist security strategy in promoting a cosmopolitan ethic. Such a military should reflect and promote cosmopolitan ideals. Hence, democratic values, the rule of law, and human rights must be respected and adhered to by the military organization in its entirety.

In the Swedish military these values have been articulated, for example, in 'the new values for the defense' (*Försvarets värdegrunder*) discussed in the previous chapter. The 'new values' have one important aim: the defense of the democratic values in society. In military training and in relation to the individual soldiers, the respect for laws against discrimination and for tolerance of difference have become particularly important.[7] Such internal organizational changes have a powerful transformative potential because, historically, the SAF, like other military organizations, has put its emphasis on the particularities of war and made the military organization virtually immune to civilian law and democratic principles. Concerns for democratic rights and values were considered a luxury, reserved for peace and organizations working for and in peace, and were not possible during times of war or high security risk.

The Swedish government does not fear a feminization of the military through peacekeeping operations. On the contrary, it seems to encourage and even embrace this development in international missions. In 2004, for the first time in the history of Swedish defense strategies, a governmental bill (Proposition 2004/05:5) emphasized gender. It stated that gender perspectives should be highlighted in international activities and that gender perspectives should be applied to all decision-making levels in international crises management (Ds 2004:30; Försvarsberedningen 2004). This represents one of the foundations for the Swedish military's image of a gender-aware, cosmopolitan-minded military.

Swedish policymakers have taken a proactive stance, for example, as one of the first countries to develop a National Action Plan (NAP) for the implementation of UN SCR 1325 in 2006. The second version of the NAP (2009–2012) set out strategies and measures that dealt with Sweden's relation to the work of regional and global security organizations. The emphasis is on including more women in peace-building tasks and in decision-making positions and paying particular attention to the effects of conflicts on women and to their role in conflict prevention, resolution, and peace building. On the national level, the NAP emphasizes education and training efforts and targets a number of organizations involved in peace and security-related activities, as well as four different ministries and the agencies involved in training and sending international staff. The SAF is among these.

Equality issues are dealt with inside the military organization through the equality strategy, the 'new values', and UN SCR 1325. According to the SAF's 2006 annual report, the military has been busy implementing UN SCR 1325 mainly through the Genderforce (2007) project and in the training of soldiers in preparation for international duty. The SAF 2007 annual report stated that about 1,700 people had completed training on UN SCR 1325. This represents a majority of the staff involved in missions abroad (SAF 2007, Appendix 3). The emphasis is on education and training at all levels of the international peacekeeping mission's organization. The view is that this will also have the greatest impact on local activities in conflict zones. To set an example, the Swedish military hosted a one-week training for gender advisors, open to military staff in European countries. This also indicates that the military is interested in depicting itself as seriously engaged in gender issues.[8] The aspirations are that through this intensive educational effort, gender issues will continue to be a natural part of military training. The SAF takes this quite seriously (SAF 2009, Appendix 3, 10–11) and shows pride in having appointed the world's first gender advisor, Charlotte Isaksson, in 2006. Isaksson became gender advisor for the European Union Force and assisted the UN Mission in the Democratic Republic of Congo, also known as MONUC (SAF 2006). Regarding these efforts to implement a gender perspective in international missions, the Swedish military seems way ahead of most other cosmopolitan militaries. I scrutinize this closer in the following chapter.

THE AMBIGUITY OF A MILITARY IN PEACEKEEPING

Associating cosmopolitan ideals with military practice has been subject to critique from various angles. Realist international relations and military strategists consider it too idealistic. Speaking from a critical international relations perspective, Vivienne Jabri (2007, 120) sees the transformation of war toward an imperative to protect other populations—'saving distant others', using the terminology of cosmopolitanism—as a conceptualization of war that sets the intervener in terms of "the protector/saviour and the subject population as the protected/saved." The concern is whether there is any possibility for agency for those to be protected, who may be 'saved' while also victimized (see also Young 2007, 117–139). Another critique comes from postcolonial perspectives. The notion of cosmopolitan militaries is viewed in the context of a new colonialism and of a *mission civilisatrice* (civilizing mission; Paris 2002). International missions are seen as missionaries of liberal democracy mainly interested in imposing democratic values on non-Westerners as a step to marketization, access, and exploitation of their economies and perhaps to secure customers for military exports.

A 2009 information pamphlet about the SAF is troubling in this respect. The document from 2009[9] is also available in English on the SAF website, where it is stated that the guide contains "all the information you need to get a good picture of what SAF are all about." The booklet is generously illustrated. An analysis of the photos shows that 17 photos depict men, mainly in uniform and 15 pictures show military equipment. Often the two are combined in one photo, showing military men using military equipment. The booklet does not mention any of the peacekeeping activities that the military is engaged in. This is very odd, since international missions have elsewhere, such as on web pages and in annual reports, been reported to be an important part of the activities of the postnational military. It is also rather perplexing to note that there is absolutely no mention of the 'the new values of the defense', and nothing about diversity, tolerance, democracy, gender, or UN SCR 1325. Instead, there is fairly detailed information about materials used, economy, and employees as well as a display of the different emblems of military ranks. The dominant iconography is of men, weapons, and military equipment.

If the booklet available is telling us everything we need to know about the Swedish military in 2009, as it claims, it is indeed another kind of military than what was previously sketched as cosmopolitan. The booklet contradicts the image of the progressive, gender-aware peacekeeping and cosmopolitan-minded Swedish military, the vision perpetuated by the SAF in other aspects of its activities. The English version most likely has been produced with an international audience in mind to communicate information on the Swedish military to the world. Many photos also contain the Swedish flag, marking a national identity. Four of the pictures combine man and material with the flag, thus projecting an image of a national, militarist, and masculine organization. The fact is that since the end of the Cold War, Sweden has steadily increased its share of the weapon sales in the world. In 2009 Sweden was ranked as eleventh among the world's producers of arms for export (SIPRI Arms Transfer Database 2009) A more schizophrenic image of the Swedish cosmopolitan military emerges as this document is taken into account.

As we have seen previously, the fact remains that, in parts and parcels, the organization is geared toward peacekeeping and international activities and thus gives rise to a view of Sweden's foreign policy as one that shows solidarity, focuses on human rights, respects the rule of law, and is democratizing itself while also demilitarizing its activities. On the other hand, it seems another part is continuing to nourish an image of a militarist Sweden, with military machinery available and ready to be used by men who are not really affected by civilian-derived norms or gender politics. The mission of the Swedish military organization thus appears highly contradictory when viewed from feminist and cosmopolitan perspectives. Perhaps it is not contradictory at all; it may be the 'destiny' of the postnational military because it must combine two conflicting and even incompatible elements of war and peace. It must be prepared to inflict violence while still being aware that peace, conflict prevention, and responsibility to protect 'distant others' is the only path to a better world.

A recent example of how these contrasting elements are embedded in the Swedish postnational defense practices was articulated in relation to the Swedish response to the UN SCR 1973 (February 2011). As a way to show solidarity with the Libyan resistance, Sweden offered to assist in maintaining the no-fly zone and contributed with around 100 people and eight JAS 39 Gripen as well as some other aircraft from the Expeditionary Air Wing.[10] While it was seen as an action in line with Sweden's cosmopolitan approach, at the same time it was quite frankly a way to show the fighter planes in action ("Ove Bring" 2011; "Prövas inte i strid nu heller" 2011) and a good way for the weapons industry to get JAS planes conflict credentials (Ek 2011; "Gripens rykte hänger i luften" 2011).

For postnational militaries to be involved in peacekeeping in line with cosmopolitan ideals requires a range of transformations that have also been exemplified here. Cosmopolitan-minded militaries have some difficult dilemmas to resolve related to how, when, and where violence can be used. Since it is argued that military skill may be necessary, the use of violence in combat situations can become necessary and thus is an option even if taken as the very last resort. If violence and combat are acceptable, albeit only under certain specific conditions, it also implies the need for weapons production of some kind, producing the tools to be used in combat. In my view, feminists who are involved with questions of war and conflict must allow space to ask whether there are times when violence might actually be necessary. Looking at violence as a possible option, not a desirable one, for example in the protection of gross violations of human rights, genocide, mass rape, and so on, does not reflect a glorification of violence or an acceptance of militarism.

From what positions can decisions be made to use violence, force, and combat in peacekeeping missions? Perhaps there is a need for guidelines for cosmopolitan militaries. As Coker (2008) and Sjoberg (2006) argue, in this postnational situation the need for an ethics of military peacekeeping is urgent. Although ambiguous in many ways, the notion of cosmopolitan militaries remains a powerful idea. Cosmopolitan militaries are already founded on the idea that cosmopolitan ethics can influence military practices. The call for ethics for military peacekeeping resonates with what Laura Sjoberg (2006) articulates on the need for an ethics of war that is feminist. Feminist ethics can be a tool to guide activities, for example, so that violence is used only when based on good political judgment. Notions of emphatic cooperation and ethics of care can help in making informed, just decisions on when to use or not to use military violence and when it is necessary to risk peacekeepers' lives to enable the respect for human rights and dignity.

An interesting link between cosmopolitan values and the ambition to save 'distant others' is made by James Pattison (2008) who argues that there is a connection between cosmopolitan views and 'the responsibility to protect,' an important norm in the global security context. The UN, and particularly the Security Council, is the most important global actor. Peacekeeping, like the ISAF studied here, is based on UN resolutions agreed to in the Security Council. Most of the peacekeeping activities to date have taken place under the auspices of the UN. The UN also enforces gender mainstreaming through, for example, UN SCR 1325, discussed earlier. Interestingly, Joan Tronto (2008) also recognizes the importance of 'the responsibility to protect' as a norm for global relations and the importance of peacekeeping in this. Tronto links peacekeeping with the feminist notion of an ethics of care. She suggests that the increased emphasis on the 'responsibility to protect' signifies a paradigm shift in global security relations. Furthermore, Tronto (2008, 181–189) argues that this shift from what previously was a 'right to intervene' to 'the responsibility to protect' is a move from a rights-based logic to an ethics of care. Thus Tronto argues that peacekeeping in connection with the responsibility to protect—or, I add, the will to save 'distant others'—can be considered care work. She defines care work as "everything we do to maintain, continue, and repair our world so that we may live in it as well as possible" (Tronto 2008, 180). Viewing peacekeeping as care work suggests that 'the ethics of care' can be used to evaluate and formulate peacekeeping practices. What is crucial in all this is to develop a sensitivity that allows the needs of 'distant others' to guide the actions and practices of the peacekeepers sent out to save them. Human rights norms are important in this, but it requires more; it requires empathy. Laura Sjoberg (2006, 49) suggests that emphatic cooperation can be the basis of a feminist international security ethic that pays attention both to care and to justice. It is a way for feminists to actively engage in debates on cosmopolitan militaries, what their tasks should be, and when violence can be used. It is particularly relevant as an ethics for peacekeeping because it provides guidelines as to how to relate to the 'distant others' that the cosmopolitan military is to protect and show solidarity with. Emphatic cooperation is based on communication and emotional identification. "Communication is the dialogical process of discovering others' needs, wants and emotions. Emotional identification is the process of attempting to identify with others' thoughts and feelings." (Sjoberg 2006, 211). This can be a valuable contribution to reform the cosmopolitan military engaged in peacekeeping activities.

Empathy is an honest willingness and an ability to appreciate the other's behavior. A suggestion on how to do this is made by Sandra Harding (1991, 268–295). She encourages us to "re-invent ourselves as others." To use empathy requires that we must carefully examine the background, that is, the social contexts that influence both our own behavior and the behavior of others. Evelyn Fox Keller (1985, 117) sees empathy as "a form of knowledge of other persons that draws explicitly on the commonalty of feelings and experiences to enrich one's understanding of another in his or her own right." Christine Sylvester (1994, 317) says that empathic cooperation is "a process of positional slippage that occurs when one listens seriously to the concerns, fears and agendas of those one is unaccustomed to hearing."

CONCLUDING REMARKS

This chapter investigated the postnational defense in terms of the relationship between security politics and the military organization involved in international missions. It noted that while there is ample evidence of a move to cosmopolitan-like military behavior in the postnational defense, the transformation takes place in the presence of a more traditionalist view that is keen to emphasize national interests, combat, and the importance of masculinity to the organization. This leaves us with an ambiguous sense as to how the postnational defense is developing.

In this chapter, the theory on cosmopolitan militaries was used to understand the contemporary military transformation in Sweden and in other midsize powers that also have embraced cosmopolitan values. The analysis focused on three challenges with the following results. First, there was ample evidence of the ambition to denationalize the military. Clearly, security policies focus on the well-being of people in distant places. However, the extent of that solidarity is challenged when violence increases and the security situation becomes worse and thus raises serious questions about the kind of sacrifices that can be expected from the cosmopolitan soldier. Second, demilitarizing the military and turning it into an organization that mainly promotes peace pointed to a core problem related to the fact that the military's unique skill and professional specialization includes the use of weapons and violence, which is not easily translated into practices needed in peacekeeping. A challenge is to develop those other skills needed while still thinking about when cosmopolitan militaries may need to use violence. To understand when, how, and where violence can be used, this chapter argued for the need of an ethics. Cosmopolitan ethics—norms like the responsibility to protect and feminist ethics based on empathy and care—can be useful for that purpose. Finally, the chapter investigated whether the attempt to democratize the military has challenged hegemonic masculinity and opened up gender aspects and tolerance for LBGT persons. For Sweden this was certainly true, as the Swedish military organization is 'marketing' itself as a gender-aware military. This is studied in more detail in the next chapter.

CHAPTER 4

Postnational Peacekeeping and the Construction of Sex and Gender

Postnational militaries are extensively engaged in missions abroad; they are often multilateral and have peace enforcement, peacekeeping, and peace building as their major tasks. The focus of this chapter is on the peacekeeping activities[1] of the postnational military. As argued in the previous chapter, some of these militaries are cosmopolitan minded. The fact that the Swedish military sees itself as a cosmopolitan, gender-aware military is analyzed from the perspective of how gender and sexuality is performed in the peacekeeping forces. This chapter connects to the previous discussion on how sexuality and gender is an organizational resource in institutions where masculinity dominates; it takes a similar approach but applies it to peacekeeping activities and asks: How are gender and sexuality organized in peacekeeping operations, and how are masculinities and femininities constructed through missions involved in peacekeeping?

The analysis begins by discussing the Swedish military's gender awareness through the example of its work in Afghanistan and the ambition to rethink masculinity. The chapter moves on to address how sexuality is related to masculinity constructs through cases of sexual misconduct by peacekeepers. Sexual exploitation and abuse undermines the legitimacy of United Nations (UN) peace operations. The chapter continues with a focus on what women peacekeepers are expected to contribute to international missions. The relational character of gender constructions is apparent here; the peacekeeping femininity can heal and remedy the 'bad' masculinity of sexual misconduct. Female presence is perceived as giving renewed legitimacy to the mission. The resource perspective discussed in chapter 2, that women are interesting because they provide something qualitatively different to the military organization, appears relevant here. Women are viewed as resources also because they are able to talk to and engage with local women. Thereby the mission can fulfill some goals of UN SCR 1325 and provide a more complete security to all. Female peacekeepers are also able to gather information. Because they are the only ones able to talk with the local women, they are crucial for gathering intelligence in highly patriarchal settings such as Afghanistan.

GENDER-AWARE SWEDISH PEACEKEEPING IN PRACTICE

With the help of a comparative study on how UN SCR 1325 has been applied in five different contingents of the International Security Assistance Force (ISAF) deployed in Afghanistan, it is possible to evaluate how gender-aware Swedish peacekeeping is in practice (Olsson and Tejpar 2009). The study's starting point is that gender equality is expected to contribute to operational effectiveness, and the analysis consists of a comparison of day-to-day military operations of the Dutch, Italian, New Zealand, Norwegian, and Swedish Provincial Reconstruction Teams (PRT). Each case is based on an average of 20 interviews on location during 2008 and 2009. The conclusion of the report is that the approach to gender equality varied greatly in the different teams. Only three—Norway, the Netherlands, and Sweden— had National Action Plans on UN SCR 1325 to guide their work. The claims that the Swedish military is gender-aware were verified: "Sweden was the only country which applied a more systematic approach and made use of both a Gender Field Adviser and a network of Gender Focal Points" and had contacts with local women's organizations (Olsson and Tejpar 2009, 115). Based on this comparative work, the report makes recommendations on a successful and systematic approach for implementing UN SCR 1325. It stresses the need to take gender issues seriously at the top level of leadership, exemplified by the appointment of a gender advisor who advises the leadership directly and oversees gender training activities. Gender field advisors are also important and should be connected to a network of gender focal points to integrate the gender activities in daily operations (Olsson and Tejpar 2009, 125). These were the strategies taken in the Swedish team. Although pointing to Swedish PRT as best in class, Olsson and Tejpar note that there is a problem of female recruitment in all five of the PRTs. The percentage of women among military staff ranges from 6 to 14 percent. It is notable that Sweden, ranking highest on gender awareness, is behind on recruitment. For Sweden, the number of women in the PRT studied was 10 percent.

In the previous chapter, it was argued that Swedish security and defense politics seem to be taking gender issues rather seriously, exemplified by the support for UN SCR 1325 and 1820 in National Action Plans. They are also taken seriously within the military organization in terms of training and in appointing gender advisors. Is the result of the study by Olsson and Tejpar (2009) suggesting a decoupling of gender and women's agency? It is Sweden's emphasis on gender awareness in operations and UN SCR 1325 training that makes them the winners in the comparative study. The authors' findings seem to suggest—although this is not their conclusion—that it is not necessarily the number of women present in the peacekeeping forces or in the military that is the key to gender awareness, but rather it is the systematic work with gender strategies, structured from the leadership level in the organization downward, that makes the difference.

Another interesting finding came out of this rich empirical study. That is that "*all* PRT personnel interviewed in connection with this study had received information about women's and men's situations in Afghanistan in the form of being told that male soldiers could not address or even look at Afghan women" (Olsson and Tejpar 2009, 121 [italics added]).

Can it be that the message is so easily heard and taken in because it appeals directly to the image of men as the protector and to a masculinity associated with chivalry? It fits nicely with what Vivienne Jabri (2007, 94–135) calls a discourse of protection that has often come to legitimize interventions. Iris Marion Young (2007, 121) calls the "courageous, responsible and virtuous man" (i.e., the peacekeeper) who protects the subordinate and vulnerable Afghan women against the patriarchal masculinity of Afghan men "the protector." The Swedish peacekeeper in the ISAF is courageous because he is risking his life for Afghan people; he is responsible because he respects the security situation and considers the consequences for the Afghan woman if he breaches the rule of traditional patriarchy; and he is virtuous because he is gender-aware. Is this also a way to construct the Afghan women as vulnerable, as victims, and then depriving them all of agency? The fact that the peacekeeping men of the PRT in Mazar-i-Sharif were all convinced they should not look at or talk to the Afghan women could be the result of cultural sensitivity training, which seems highly important. Yet Lutz et al. (2009) point out that the cultural training peacekeepers receive often contains cultural stereotypes, sometimes bordering on racism. This is a complex issue. Palwasha Hassan (2010) of the Afghan Women's Network pointed to a problem associated with what she defined as a "cultural oversensitivity." It tends to stereotype Afghan women's role in society as static and fixed. However, respect for the modesty of women in Muslim culture appears to have prevented sexual exploitation and abuse in United Nations Interim Force in Lebanon[2] (Lutz et al. 2009, 13–15).

While gender, women, and UN SCR 1325 are by now—at least to Swedish contingents—familiar concepts, other related concepts such as masculinity or sexuality are not part of the peacekeeping colloquial. The masculinity of the patriarchal Afghan leaders and the bad masculinity of the Taliban warriors is taken to be a given or only anxiously treated in the frequent encounters of daily military operations. An example of this anxious, careful approach to 'the other' masculinity is narrated by a Swedish female peacekeeper. She went with her male peacekeeping collegue to visit a village and beforehand was advised by him about the local culture. He said that she should not offer her hand to the Afghan men but instead simply nod, place her hand on her heart, and then stay in the background. He would do the talking. When they entered the village, the Afghan men were very hospitable; they enthusiastically shook hands with her and patted her shoulder. When it came to discussing their business, the group was very interested in hearing what she had to say. This experience made her male colleague realize that a female officer can actually contribute by relaxing the atmosphere. This was the lesson learned (i.e., that a woman's presence can contribute positively to the atmosphere of village-level negotiations; see Ivarsson and Edmark 2007, 23). Another interpretation would be that her colleague acted as the masculine protector and took on the responsibility of protecting her from the 'other' masculinity, which he was nervous and unsure about. Her presence, as a Western woman in uniform, was a curiosity and neutralized, for the moment, the tension of the different 'masculinities' in the room. Furthermore, the account suggests that the 'cultural oversensitivity' that Hassan spoke about relates to Afghan men and their masculinity as well.

In the gender-aware Swedish cosmopolitan military there is space for thinking differently about masculinity, and some Swedish peacekeepers seem aware of the importance of conveying an image of a new and different masculinity as related in interviews (see Ivarsson and Edmark 2007). In speaking about male peacekeepers, many female officers emphasized the importance of the male peacekeeper adopting alternative gender roles as examples of alternative images of masculinity. These informants said that the Swedish male peacekeepers often exhibited a different masculinity, which was also conveyed to the locals. One example occurred when the troops based in Kosovo made regular visits to orphanages. The peacekeepers played with and showed an interest in the children at the orphanage. Simultaneously, they provided an alternative image of an activity—an activity that the group of peacekeepers saw as valuable. This way they also projected an alternative picture of masculinity and showed that a man, too, can care for, enjoy, and value playing with children. Another example was that of a male military officer, in uniform and cooking dinner for himself. While he was cooking the locals could see him through the open door in the kitchen. Thereby this military officer was seen doing something perceived in the local context as a solely feminine task. This challenged traditional gender roles as perceived by the local community (Ivarsson and Edmark 2007, 29) at the same time it conveyed and projected an alternative image of masculinity.

Sweden's cosmopolitan-minded military has conducted various gender mainstreaming efforts, including systematic organization of training in UN SCR 1325 and gender issues. The peacekeeping missions also open up opportunities to articulate 'alternative' masculinities, and both the men and the women of the mission can become models of an alternative gender order where they are deployed. Yet the task of questioning or reformulating masculinity was not part of a strategic or collective effort of the organiza-tion.[3] Processes of identity construction were imbued with cultural under-standings and stereotypes.

PEACEKEEPERS' MASCULINITY AND SEXUAL MISCONDUCT

The construction of masculinity in relation to sexuality is an important aspect of peacekeeping missions and their relation to the local community, but it is not approached in a direct way. Sexual abuse and rape seem always to have been part of armed conflict and war. To that end, it has been treated as a nor-mality of war. This does not mean that rape in war is a universal phenomenon or that it has to be this way (Goldstein 2001, 362). However, it is only during the past few decades that rape and sexual abuse have come to be viewed as elements of conscious conflict and war strategies, rather than as unfortunate side effects of war and violent conflicts (Diken and Laustsen 2005; L. Skjels-baek 2001; Stiglmayer 1994). Ruth Seifeirt (1996, 35) writes that: "Mass rapes have occurred in all modern wars, but not until the gender-specific atrocities committed in Bosnia-Herzegovina have they attracted worldwide attention." It has also become a central part of security and conflict analysis (Hansen 2001; Mukwege 2009; Münkler 2005, 81–87) and has brought the issue of sexual violence against women and girls to the global agenda (Farwell 2004) of the International Criminal Court and the UN with SCR 1325.

The urgency of the issue was confirmed in June 2008 when the Security Council adopted Resolution 1820 and with it "demanded the immediate and complete cessation by all parties to armed conflict of all acts of sexual violence against civilians." It also expressed a deep concern regarding the finding that, "despite repeated condemnation" through other UN work and in previous resolutions, "violence and sexual abuse of women and children trapped in war zones was not only continuing, but, in some cases, had become so widespread and systematic as to reach appalling levels of brutality" ("Security Council Demands Immediate and Complete Halt" 2008). Resolutions 1888 in 2009 and 1960 in 2010, as well as UN Secretary General Ban Ki-moon's appointment of Margot Wallström as the special representative on sexual violence in conflict, attest to the increased urgency of the problem and the salience of it in the UN context.

Since rape, sexual violence, and abuse are widespread in current conflicts, it is highly relevant to investigate sexuality and its conceptualization in regard to military engagement in peacekeeping activities. Peacekeepers are sent out to protect locals from human rights abuses and help them resolve conflicts, ultimately creating the conditions for peace. Any type of abuse, sexual or other, undermines these efforts and threatens the legitimacy attached to peacekeeping missions (L. Anderson 2010; Grady 2010). Unfortunately, it is all too common. The most well-documented cases of wide sexual misconduct concern the United Nations Organization Stabilization Mission in the Democratic Republic of Congo, or MONUC (Higate 2004). A UN report states: "Interviews with Congolese women and girls confirmed that sexual contact with peacekeepers occurred with regularity, usually in exchange for food or small sums of money. Many of these contacts involved girls under the age of 18, with some as young as 13" (United Nations General Assembly 2005).

Despite the fact that by 2003 the UN Secretary General had already instituted a zero-tolerance policy against 'the blue helmets' sexual relations with locals, it continues to be a problem. Many different nationalities are involved (Hull et al. 2009, 42; Lutz et al. 2009). A television documentary from 2010 verified that sexual relations and prostitution continues in the Democratic Republic of Congo and that Swedish staff have also been involved in this. There seems to be a broad tolerance for sexual relations of all kinds through the UN peacekeeping organization, and impunity prevails (Yllner 2010). Because the main goal of a peacekeeping mission is to create trust in the operation among locals, sexual misconduct threatens and will sever that relationship.

Higate and Henry (2004) argue that sexual misconduct is connected to how masculinity is constructed within the military organization that provides the staff for peacekeeping operations. According to the authors, male peacekeepers' masculinity is reconfirmed in relation to local women, and sexuality often continues to be constructed in line with the traditional understanding of men's sex drive. Men's sexuality is often understood in terms of men's 'need' of sex. Typically, men's sexuality is perceived as a natural, instinctual, and largely uncontrollable phenomenon. Lutz et al. (2009, 7) speak about the UN's "hydraulic model" of male sexuality: Male sexuality builds up and has to be released. It is similar to the traditional understanding of rape in war (i.e., as an unfortunate but 'natural' thing). Such an understanding assumes that men cannot control their own sexuality. Men are assumed to lack agency when their sexuality is constructed in a way that simply presumes it to be about a biologically generated 'drive'. Often associated with this is an understanding of women as passive sexual objects and/or women as the property of men (Jennings and Nikolic-Ristanovic 2009, 12).

Paul Higate (2007) calls attention to the way masculinity was understood in the context of peacekeeping forces active in the Democratic Republic of Congo and Sierra Leone. Masculinity was conceived as simultaneously both strong and weak. Based on interviews, Higate (2007, 106) concludes that, in general, "the male sex drive was seen to reflect an integral component of deep masculinity." This was a view beyond question, endorsed by both peacekeepers and the locals. Thus local women and girls provided sexual favors 'needed' for the strong masculinity in terms of the sex drive. At the same time, the economic desperation of many local women and their families gave rise to the construction of a weak masculinity that could not withstand and was "vulnerable to the predatory approaches of women and girls" (Higate 2007, 107). It was a weak masculinity because male peacekeepers felt like victims of women soliciting sex. Lutz et al. (2009, 6) describe this happening among peacekeepers stationed in Haiti, where women in mission areas "throw themselves at peacekeepers." Again, masculinity and male sexuality becomes something out of the hands of the male peacekeeper, who could not possibly be responsible for an innate biological drive or for the lecherous solicitation of sex by young local women. Sexual desire, as articulated by the men involved, becomes constructed also as something outside a man's own agency.

Yet when male peacekeepers were asked to tell their stories, they seldom articulated it as a natural sex drive; instead they preferred to tell stories of sexual liaisons with local women and girls as "carefree, sexual and romantic" (Higate and Henry 2004, 491). This was also the picture that emerged from Swedish peacekeeping in Congo in the television documentary referred to previously. Elise Barth (2004) noted something similar in the UNMEE forces based in Eritrea (United Nations Peacekeeping Department 2008a). The peacekeepers that Elsie Barth interviewed felt they were helping the girls when they gave them money. Although the soldiers had codes of conduct to follow, it was unclear to them where to draw the line between prostitution and relations that they perceived to be of mutual consent. Yet there were cases of sexual abuse; for example, Irish soldiers made pornographic movies using Eritrean girls, and Danish soldiers were convicted of having had sex with a 13-year-old Eritrean girl (Barth 2004, 13–14; cf. Hull et al. 2009, 40). Although it has been documented (e.g., Otto 2007) that there are a range of different types of relations between locals and peacekeepers, and some may very well be romantic and based on true love, most are probably not. However, male peacekeepers do not tell their stories as those of sexual misconduct and abuse; neither do they tell of the deep asymmetries of power and privilege that clearly exist between peacekeepers and locals in any peacekeeping setting. The peacekeepers in Eritrea did not see themselves as exploiting the women, even though "compared to the Eritreans, the soldiers are rich" (Barth 2004, 15–16).

From the perspective of poor local women and girls, such sexual relations may look less romantic. Sexual and other relations with peacekeepers can offer a venue to a different life and many times a way to earn an income to support themselves and their families. They perform survival sex. This brings some nuance to a dominant view that often comes across in UN and other documents on sexual abuse in conflict. It is a view that constructs women and girls' vulnerability in terms of a complete lack of agency facing a predatory military, an inaccurate portrayal. Women have some agency, although often very circumscribed; they have made a choice to sell sex for their and their family's survival. Nevertheless, taken into consideration the unequal power relations between peacekeepers and the local community, sexual liaisons and prostitution often look a lot like exploitation. It was clearly the case in Congo and in Sierra Leone (Higate 2007). Furthermore, gendered and racialized stereotypes are frequently combined. Jennings and Nikolic-Ristanovic (2009, 6–7) found evidence of this in the context of Haiti and Liberia. Peacekeepers identified those societies as normally promiscuous and of the women as subservient. Therefore, they could justify having sexual relations with locals for money and food, because the women were perceived to be 'easy' and think differently 'about sex'. Haiti and Liberia are countries with a permissive state policy on sex work. Due to this, the peacekeepers, including their superiors, found it difficult to reinforce the zero-tolerance policy. The argument was they could not be expected to behave different toward local women than the local men did (Jennings and Nikolic-Ristanovic 2009). Rather, it seems as if the peacekeeping missions' organization and leadership is lax about the codes of conduct and are unwilling to follow and enforce the rules and regulations that apply, such as the UN codes of conduct (Lutz et al. 2009).

If the leadership of peacekeeping missions understands men's sexuality in the traditional interpretation as a natural and biological need that should be tapped and controlled within the organization but must have an outlet, then this may explain the leaders' willingness to accept sexual misconduct in spite of UN policies. This interpretation of men's sexuality tends to individualize the problem of sexual misconduct and thereby conveniently exclude the role of the peacekeeping organization, the military, and other societal organizations. It is well known through Cynthia Enloe's (1993, 2000) work, for example, that military organizations have problematic relations to heterosexual men and their sexuality in part because militaries have used sexuality as an organizational resource. For example, the military organization's emphasis on group loyalty or unit cohesion is "a significant factor in creating a wall of silence around misconduct" (Lutz et al. 2009, 9). Chapter 2 discussed the importance of enforcing heterosexuality to achieve homosociality for group cohesion and military effectiveness. Here, homosociality and group loyalty can become a part of fostering a culture of impunity around sexual abuse by peacekeepers.

Regarding the UN's stance on peacekeepers and their sexuality, Jennings and Nikolic-Ristanovic (2009, 20) write that the view of male sexuality that comes across in UN actions and policies illustrates that there is an "unwillingness on the part of the organization to see itself as part of the problem"—after all they have a zero-tolerance policy. The view projected through the UN's policies and actions is that peacekeepers' sexual misconduct is an unfortunate and exceptional act, the acts of a few delinquents—a few bad apples—among the peacekeepers. The solution to the problem is seen as coming through the fine-tuning of organizational rules and sanctions. However, a failure to address the understanding of men's sexuality that is embedded in the institutions involved means the problem of sexual misconduct cannot be 'resolved'. The UN's message on sexual misconduct is double-edged: it is concerned about the gravity of the practices and wants to take stringent action against them, but it also wants to make sure that the conduct is not seen as too general of a phenomenon as to damage UN peacekeeping legitimacy (Kanetake 2010, 209).

ORGANIZATIONAL ASPECTS ON SEXUALITY IN PEACEKEEPING

As evidence shows, there are recurring incidents of sexual misconduct despite the UN zero-tolerance policy that has been in place since 2003. Cynthia Enloe (1993, 2000) and Katherine Moon (1997) have stressed the connection between militarism and sexual exploitation as a factor influencing the emergence of sex economies. Enloe (2000, 51) writes: "Ideologies of militarism and sexuality have shaped the social order of military base towns and the lives of women in those towns." She considers this highly relevant for peacekeeping as well. Where militaries and international missions are stationed, peacekeeping economies develop (Enloe 2000, 91). This is the experience from missions in the past, such as those in Bosnia, Kosovo, Haiti, Cambodia, and Liberia. Kathleen Jennings (2010) argues that it is a common economic phenomenon that develops when a peacekeeping operation arrives in an area. Peacekeeping economies develop due to the resources that the missions carry with them, their need for local staff to assist them, and the difference in income between peacekeepers and the locals. Sexual relations can become a part of and a resource in this, explored by local women and men as a way to earn extra income. The sex economy (sex work, trafficking, sex tourism) depends on locals being sexually available for peacekeepers and other international actors. When locals are not available or cannot fully cater to the demands, trafficking becomes an alternative. Thus, sex industries are often an important element of the peacekeeping economies that are, it should be noted, "highly gendered—but . . . the 'normalization' of peacekeeping economies allows these gendered effects to be overlooked or obscured" (Jennings and Nikolic-Ristanovic 2009, 2). A failure to recognize this undermines and counteracts the efforts to gender-mainstream peacekeeping operations.

Experiences from Bosnia and Kosovo show that the sex industry continues its operations after peacekeepers leave as, for example, sex tourism. The effects are often lasting and spread throughout the region (Jennings 2010; Jennings and Nikolic-Ristanovic 2009, 15). An economic, ethnic, or postcolonial view needs to be added to the analysis of sexual misconduct of peacekeepers and the sex work of locals (cf. Agathangelou and Ling 2003). It must be recognized that survival sex offered by locals "generates income or leads to access of privileges and resources that are necessary for everyday material survival" (Otto 2007, 260). Dianne Otto argues that the UN's approach to sexuality and peacekeeping ignores and thereby also perpetuates the complexity of sex economies and that the cultural and economic inequalities between peacekeepers and the local population are silenced. By ignoring this element of peacekeeping missions and the economies they generate, the UN fails to address the underlying questions relating to social justice and inequality. The impact of the peacekeeping missions is, in this respect, neither localized nor temporary. Often the legacy of the peacekeepers' presence remains and develops into a sex tourist site, as in Cambodia, or sex trafficking, as in Bosnia. Kathleen Jennings (2010, 231) does say, however, "Afghanistan seems to be the most obvious exception to this generalization."

For the current situation in northern Afghanistan, where the Swedish ISAF are deployed, it may be taken as a given that no sexual relations or fraternization with locals are possible, first because, as shown earlier in this chapter, the staff in the PRT teams received the message loud and clear to avoid addressing or looking at Afghan women. The second reason is that the local gender regime dictates a very strict control of local women, although evidence suggests that this may not be true (Dixon 2010; Nasuti 2009). Prostitution was sanctioned by the regime during the Taliban rule ("Prostitution under the Rule of Taliban" 1999), and there is evidence of an emerging sex economy in Afghanistan. This also includes trafficking, particularly of Chinese girls. By now, it is widely known that young Chinese women work out of Chinese restaurants in Kabul and cater to the international community that is based there (Huggler 2006). The sex economy also involves local women (Zimmerman 2008). "While Afghanistan's strict Islamic law forbids prostitution, there are signs the work is taking formal root, with brothels operating in some cities and pimps managing prostitutes. Bribes take care of unwanted police attention" (Qadiry 2008). Despite severe punishment if caught, women (and possibly some men) in Afghanistan take part in the sexual economy just as they do in many other impoverished places of the world. Women who have lost or been ostracized by their families may be particularly prone to sex work, since other jobs are rarely available to them (*Mänskliga rättigheter i Afghanistan* 2007, 4; U.S. Department of State 2009). They engage in survival sex. Qadiry (2008) tells of two young female sex workers in the area of Mazar-i-Sharif, in the vicinity of where the Swedish peacekeepers are based. In both cases the women do it to support their families.

While it is clear that a sexual economy and prostitution are developing in Afghanistan, including in the area where the Swedish ISAF mission is stationed, this does not mean that Swedish peacekeepers necessarily engage in prostitution or sexual relations with locals. The Swedish defense bill of 2005 (Prop 2004/05:5) recognizes that the gender approach may profoundly challenge the military units' traditional behavior but also simultaneously improve the relationship between peacekeeping units and the local population. The bill stresses, for example, that prostitution and trafficking around international missions is a violation of UN SCR 1325 and thus is not acceptable. In an effort to curb this kind of behavior among Swedish peacekeepers, the government proposed a "Code of Conduct" applicable to all Swedish citizens on all types of international missions and based on Swedish legal codes. To buy sex is a criminal act, as are discrimination, abuse of power positions, and any type of sexual abuse and violence. This law was introduced in the domestic context in 1999 and is applicable outside the country as well; it is also the base on which Swedish peacekeepers using prostitution are prosecuted.

A factor that influences peacekeepers and their officers' attitudes toward sexual conduct is the way that sex work and sexuality is viewed nationally. Attitudes about sexual relations vary among countries taking part in multilateral missions. For example, the Netherlands and Sweden have very different views on the topic of prostitution as well as on policies of conduct; thus there are differences in levels of tolerance of sexual relations and fraternization among nations. National policies affect how sexual conduct is viewed when on international missions, although the UN's zero-tolerance policy is supposed to override national policy. Part of the problem is that the UN bears only partial responsibility for the realization of its own zero-tolerance policy. The states that provide peacekeepers as well as troops also have to be equally committed to the zero-tolerance policy (Kanetake 2010).

Furthermore, policies on conduct have a norm function and influence the understanding of what is considered appropriate sexual or love relations. The UN policy that the peacekeepers must conform to discourages having sexual relations with the local population, but it also forbids its personnel from having sex with anyone under 18 and from buying sex from prostitutes (United Nations Secretariat 2003). Because it forbids prostitution and particularly because it sets the age of consent at 18, it is considered a very strict policy in relation to the national policies of countries providing peacekeepers. Dianne Otto (2007) is highly critical of the Secretary General's zero-tolerance policy, because, she says, it exemplifies a highly conservative understanding of sexuality, one that is matched only by national legislation in a few very sexually conservative countries. In comparison, the Swedish code of conduct reflects the restrictive position on prostitution or sex work in Swedish society. According to Swedish law (enacted in 1999), it is illegal to buy sex; this is one of the more radical legislations against prostitution in the world. The law criminalizes the buyer of sex and any organized solicitation of sex (C. Ritter 2008). It is also valid outside the country's borders. The Swedish code of conduct is thus similar to the UN zero-tolerance policy, although it differs on the age limit for consensual sex. In Sweden the minimum age of consent is 15, whereas the UN sets it at 18. In this respect, the UN policy is more restrictive than Sweden's and many other countries' around the world.[4]

Otto (2007) points out that the sexual conservatism constructed through the UN zero-tolerance policy goes further to make the only acceptable relation between a local and a peacekeeper a heterosexual marriage. This is not only a conservative understanding of sexuality but also a form of "sexual negativity," she argues. In an attempt to deal with impunity on sexual abuse, the policies that emerge also generate norms on sexuality. Extreme control of sexual relations is exemplified by the U.S. military deployed in Iraq and is perhaps the most extreme way by which to also construct a deeply negative view on sexuality (Kramer 2010, 365). Sexual negativity contradicts attempts to enhance the understanding of sexuality and to, for example, increase acceptance for lesbian, gay, bisexual, and transgender (LGBT) people. Just as in militaries in general, homosexuality in peacekeeping forces is basically a silenced issue. One reason for this is that homosexuality is defined completely outside UN's understanding of sexuality. Otto (2007) argues that through the zero-tolerance policies, the heterosexual male also becomes the dominant norm for the peacekeeper. A way to deal with this is to differentiate between rules on sexual conduct and fraternization within missions as an operational matter and sexual preference as a private concern (cf. Kramer 2010).

Tolerance and acceptance for differences in sexual preference have been difficult to deal with in the Swedish military organization at home, as pointed out earlier. It is likely to be more problematic in peacekeeping forces due to clashes between different value systems that come to the fore in international peacekeeping operations. Alan Ryan (2004, 66) notes that "while there is great value in asserting and pursuing a clear set of cosmopolitan objectives when committing troops to cosmopolitan operations, it also needs to be recognized that liberal Western commitment to cosmopolitan democratic values is not always matched by similar value systems elsewhere." This is particularly so when we look at more specific elements of democratic values and the extent of human rights, such as those involving sexuality. Although the Swedish military has made a move to openly discuss and aim for sexual tolerance, it is a vision and an objective that cannot easily be implemented in the peacekeeping forces that are often also multinational. As mentioned, different gender orders, exemplified by the understanding of sexuality and sexual appropriateness, are represented within multilateral peacekeeping forces. Views can be very different among nationalities within multinational peacekeeping forces but also in relation to the view of sexuality in the local context where the forces are deployed (cf. Levy 2007, 187).

Homosexuality is not widely tolerated. It is even criminalized in many countries, and in some countries homosexuals face death penalties. The question for a cosmopolitan military is how to cooperate with military personnel and civilians who have radically diverging values and views on sexual rights. Even among the fairly like-minded countries within the EU, there are important differences, and EU countries are supposed to cooperate in peacekeeping efforts. For example, in Italy, Greece, Hungary, and

Croatia, there is no tolerance for sexual diversity in the military, whereas countries such as Belgium and France have a somewhat relaxed attitude, and the Netherlands has the most tolerant policy (Parliamentary Memo 2004). For LGBT military personnel in peacekeeping forces, these different attitudes and policies make them possible targets for harassment. This was recognized in the Swedish government's budget bill of 2007. It points to the problem that LGBT persons can encounter in international operations under other countries' command. According to Proposition 2006/07:1, "Since in many countries there is institutional discrimination or legal limitations on homosexuality, the Armed Forces have to secure that Swedish soldiers are not discriminated against during multilateral operations abroad" (61). Having LGBT peacekeepers in the mission may present a security problem.

This issue is relevant for the Swedish international peacekeeping force because it must assure protection from harassment for its own staff. It can be very difficult or even impossible to ensure tolerance and acceptance for LGBT persons in peacekeeping missions if the local understanding of sexuality does not allow for this tolerance. This is the case in Afghanistan. However, in the many reports and material that I have collected, it does not come up as a problem. I can only speculate why: It may be that the issue is silenced, or it could be, as Bateman and Dalvi (2004) show in their somewhat dated but still relevant study, that to be a gay male peacekeeper in multilateral operations is not really a problem or concern because there are much more pressing issues that take priority over the sex preferences of individuals. For the cosmopolitan-minded military, different values are positioned against one another, and it may be necessary to take a stand on what is the prioritized concern—is that to gain the trust of the local community for the mission, or is it to uphold norms of sexual freedom and rights? Indeed, this puts the cosmopolitan military to the test.

WOMEN BRING LEGITIMACY TO THE PEACEKEEPING MISSION

In light of the experience with sexual misconduct among UN peacekeepers and what has previously been argued, it is possible to understand why the importance of women peacekeepers and their ability to create trust in the mission has been highlighted lately. Women peacekeepers are thought to improve the mission's legitimacy in the eyes of the local population. This is crucial. Locals continuously consider and evaluate the peacekeepers' use of authority and their general behavior in providing the security that should mean stability and eventually peace. Although not explicitly part of the five peacekeeping principles (Hansen et al. 2001, 3), it is likely that the level of trust that locals have of peacekeepers impacts the success of the mission (Mersiades 2005). One of the challenges for peacekeeping missions is to help the local population understand why certain actions are taken (Pouligny 2006). Rubinstein et al. (2008, 545) argue that this "hinges on peacekeepers' ability to interact with local peoples in ways that communicate genuine partnership and respect." The local population makes judgments of peacekeepers that they encounter, and both Mehler (2008) and Pouligny (2006) point to the need to further investigate and explore the role of peacekeepers in the local social fabric. Judgments can be positive or negative, and it is obvious "that these judgments influence the chances of peacekeeping missions being successful" (Mehler 2008, 55).

Sexual misconduct like that documented in many past missions has undermined local populations' trust of peacekeeping troops. The United Nations Peacekeeping Department (2008b, 36–37) recognizes that a high legitimacy for peacekeeping forces contributes to their success. Sexual exploitation and abuse is a threat to both their legitimacy and their effectiveness. Arguments for the inclusion of women in peacekeeping forces often articulate a hope that women can increase the legitimacy of the troops. This is one of the main arguments given by the Swedish military for including women in peacekeeping operations. A report on the unintended consequences of peacekeeping operations with a particular focus on the Swedish military has a similar approach; that is, according to Hull et al. (2009, 53), the presence of female peacekeepers is crucial, and one of their most important recommendations in dealing with sexual misconduct and abuse is to include more women.

Based on an empirical study of Australian peacekeepers, Bridges and Horsfalls (2009, 120) confirm that "female personnel assist in engendering trust, allaying fears, improving the reputation of peacekeepers, normalizing the presence of troops, and positively facilitating the peace process." One of the most important roles of women in peacekeeping is to temper inappropriate behavior. They also introduce a different kind of culture, which can install trust in the host nation for the presence of peacekeepers and thus make them more effective. The conclusion in Bridges and Horsfalls (2009) study is unique in that they make an explicit connection between the history of sexual abuse and misconduct and the view of troops with females as more trustworthy. Having said this, their conclusion seems slightly problematic. When male peacekeepers have been portrayed as perpetrators of abuse, women are to come in and "clean up the mess," to install trust and legitimacy toward the local population. Oliviera Simic (2010) is critical of this approach taken on by the UN. She sees it as a way to encourage women to "join peacekeeping operations as sexual violence problem-solving forces" (Simic 2010, 188). Underpinning the call that women should help increase the legitimacy of the UN operations by their presence is the idea that female peacekeepers, through their pacifying and nurturing presence, will discipline male peacekeepers into proper conduct toward locals (Simic 2010, 189).

When the efforts of UN peace operations turn to women peacekeepers with the hope that they will save their tarnished reputation and restore confidence and legitimacy to the operations, it may be both an unfair and all-too-difficult task. The way that femininity is constructed here is in terms of 'saving angels' or 'beautiful souls' that rescue men from their uncontrollable sexuality and militarized masculinity. Also, Swedish female peacekeepers are, in the view of the Swedish Armed Forces (SAF), expected to bring renewed legitimacy for the mission, and (re-)install the trust for the mission in the local population. The Cecilia Hull et al. (2009) study does, however, recognize that there is contradictory evidence as to what role female peacekeepers play in mitigating sexual exploitation and abuse among their own contingent. Yet the authors insist on the particular importance of women as part of the peacekeeping operations so that local female victims who have experienced sexual violation have someone they can trust and turn to. Again, there is a specific burden put on women here; Hull et al. (2009, 52) suggests having "female soldiers participate in peace support operations to deal with the impunity that often surrounds SEA [sexual exploitation and abuse] acts."

The expectations put on women are indeed high and, when placed within the stark reality of a peacekeeping operation, represent a daunting task. Try to imagine the setting in which these female peacekeepers operate. While there has been a very slow increase in the number of female military staff serving in peacekeeping, women have never comprised more than 2 percent of the total (Simic 2010, 188).[5] Johanna Valenius (2007a, 517) writes that: "Peacekeepers' camps are masculine spaces. Men dominate and define the space by their sheer numbers and uniform bodily presence. More precisely, it is not exactly *men* who characterize the space, but there is a certain type of *masculinity* that permeates and defines it, the heteromasculinity of 22- to 24-year-old-men" (italics added). In these spaces, gender is produced and performed; peacekeepers' security practices produce gender (Higate and Henry 2010, 37). Because young men take up peacekeeping spaces, their gender subjectivity completely dominates. In that space, is it realistic to expect that a few women should monitor their colleagues and report them while also fending off unwelcomed sexual invites and dealing with sexual harassment?

Furthermore, feminist research on militaries in general suggests that female military recruits are likely to try to fit into the military rather than criticize it. Although some women become radicalized in the setting and choose to be critical, most women are likely to turn their heads the other way when they suspect that their male colleagues are, for example, soliciting local prostitutes. Many actually avoid the company of male peacekeepers when they have leisure time because they do not want to take part in barhopping and drinking or in fraternizing with local women. Female peacekeepers are often careful because they want a good reputation. Their behavior and sexuality is under the constant surveillance of a disciplining male gaze. Swedish female peacekeepers tell of such experiences in the study by Ivarsson and Edmark (2007, 24–26). They must be more careful in their contacts with men but also when drinking and going out, because such actions can damage their reputation. Furthermore, sexual harassment is as frequent in the international missions as it is at home; according to the 2005 SAF general survey on sexual harassment, it may even be more extensive, as suggested by another survey directed toward military staff who had leadership positions in international missions (Ivarsson and Edmark 2007, 64).

ARE WOMEN BETTER PEACEKEEPERS?

In addition to the view that women are expected to bring legitimacy to international missions, another tendency is to perceive women as equipped with special and unique skills. There is a lot of ambiguity in how the contribution that women make is viewed. Gender difference has, in the past, been articulated within the Swedish military in terms of a resource that can be used to ameliorate military performance (Ds 2004:30, 11–14).[6] In peace operations, women have been viewed as those who have the skills needed to secure good relations and solve conflicts with the local and civilian population, because, for example, they are better communicators and equipped with the ability to calm aggression. Women are often perceived as softer, more peaceful, and more prone to cooperation (DeGroot 2001, 37; Rhen and Sirleaf 2002, 63; I. Skjelsbaek 2007, 20). This is in line with a traditional view of women as associated with peace and more likely to promote peace (Ruddick 1989).[7] Women as peacekeepers have been stereotyped with characteristics such as a "tendency for discussion, negotiation, and compromise" and associated with the "avoidance of more aggressive alternatives" (Fox 2001, 17). Hence, femininity is often articulated as being more cooperative, deliberative, and prone to peaceful solutions. When such behavior is taken to be exclusively associated with women's special nature and this is why they are considered a resource, it tends to affirm an essentialist view of gender. In the international peacekeeping discourse in general, women's contribution as peacekeepers tend toward an essentialist construction of femininity (Charlesworth 2008; Puechguirbal 2010, 181).

Rather than disassembling typecasts of gender characteristics, this view runs the risk of playing into gender traditionalism (D. Smith 2001, 44). It is ironic that the female stereotypes that have been seen as barriers to women's participation in the military in the past make them effective as peacekeepers (DeGroot 2001, 24, 34) and thus an attractive resource for the Swedish military today. The tendency to consider femininity a resource for the postnational military can be interpreted as a sign of this paradigm shift and, more specifically, as associated with dissolving the military-civilian boundaries. This shift implies that the skills and attributes historically connected to women and the civilian sphere and of slight or no interest to the national military have become the skills and attributes needed in the postnational, cosmopolitan-minded military, necessary for making peace, and thus highly valued and sought after (see also Penttinen 2011). Indeed, there is a contradiction in this, of women being both "idealized and undervalued" at the same time (Willett 2010, 143). On the other hand, equating gender to certain individual features of women does not question the gendered norms traditionally associated with military institutions. The argument that gender is a social construct and gender difference the result of historic socializing processes is excluded from that interpretation (cf. Väyrynen 2004, 137).

When looking more specifically at the material on Swedish female peacekeepers in the ISAF in Afghanistan, the view of female peacekeepers does not fully match this pattern. Female peacekeepers are expected to provide specific resources; however, the resources that women are thought to contribute to the ISAF are related to their bodies and their appearance, coupled with the expectations generated among the locals about them as female peacekeepers. It has little to do with possessing different skills, or at least it is not something that is stressed. In the SAF bimonthly newsletter (*Insats & Försvar* 2006 no. 4, 24–26), a case is made that the only way toward sustainable peace is through an emphasis on women. It is important for peacekeeping units to secure the engagement of local women in the peace process and cater to local women's needs. To accomplish this, it is argued, it is necessary to have women in the peacekeeping units (*Insats & Försvar* 2006 No. 4, 19).

Female peacekeepers become a resource to the military because it makes possible the engagement of the military with Afghan women. Female peacekeepers are important and sometimes provide the only channel to talk to, engage with, and obtain information from local women. It is only through the female peacekeepers that the ISAF mission can reach out to the entire society. Recall that the male peacekeepers had been told not to address or look at the Afghan women. Local women, who are subject to peacekeeping efforts, more readily cooperate with female soldiers and police in international peacekeeping forces. The oppressive patriarchal regime in the context of Afghanistan means that some local women's agency in the public sphere is closely checked and restricted to contacts with other women. The only way to include local women in the peacekeeping process and thus live up to the spirit of UN SCR 1325 and contribute to a more inclusive security is to engage female peacekeepers. The alternative way to communicate with and obtain any information from the local civilian women would be through men related to them. In this context, it is appropriate to point out that the UN SCR 1325, despite its many flaws, has provided women agency in global security matters, while it still remains the military peacekeeping organization's decision whether to include women and on what terms.

As a way to fully benefit from the resources that female peacekeepers can contribute, a special military observation unit was set up within the ISAF in Afghanistan in the spring of 2006. The Military Observation Team (MOT) Juliette was unique, as it was the first Swedish unit to consist of only women. It had as its one and only task to interact with local women.[8] Something similar had been tried informally a few years earlier in Kosovo, with some success. MOT Juliette exemplifies how a military can use gender difference as an integral part of the military organization's strategy in peace enforcement. Through the work of MOT Juliette, Afghan women were able take part in building security in their society. UN Secretary General Ban Ki-moon (UN Security Council 2008, 4) talked about another example similar to MOT Juliette, the all-female Indian civil police unit deployed in Liberia. He saw this as a possible model and "as an excellent example of the unique contribution that female personnel can make." He continued, "Through their sheer presence, the members of that Indian contingent were showing Liberian women that they, too, could play a role." When MOT Juliette interacted with the local community, it also increased awareness of the possibility of a different role for women. An alternative femininity was projected through MOT Juliette. There was thus some type of power associated with the female presence. The members of MOT Juliette showed that women could be peacekeepers too and that a different gender order is possible ("Hon riskerar livet varje dag" 2006). Perhaps a way to encourage societal and institutional change and

Figure 4.1: MOT Juliette on a visit to Balkh University, Mazar-i-Sharif. Swedish Armed Forces. Photo: Andreas Karlsson.

promote the transition to a humane system of governance is also a task of a cosmopolitan military (cf. Elliot 2004, 25).

In a document issued by the SAF on gender and operational effect (SAF Information leaflet, 2009) it is argued that "a mixed force consisting of both men and women will achieve greater success when gathering information and creating trust for the aim of the operation." There is a notion that cooperation will build legitimacy when local women are also able to take part, voice their opinion, and be heard by the international forces. Talking to the local people, both women and men, is part of creating a more complete security. There are two sides to this: On the one hand, it is about increasing local women's participation, for example, through the building of networks that include women, as Madeleine Jufors (2008), gender advisor for the ISAF argued, it becomes clearer how security can be strengthen. On the other hand, it is also a way to obtain information that is needed for operational effectiveness, for example information on where weapons are hidden. In gathering intelligence, women peacekeepers become a real resource to the military in the case of the ISAF in Afghanistan. An important way to gather intelligence information, such as who carries or hides weapons and where ammunition and explosives are hidden, is through the contact with local Afghan women who can provide this kind of information to female peacekeepers. This rather functional approach to integrating women in peacekeeping efforts on the ground fits well with an organization that is oriented to perform specific tasks in an efficient way. If women have the specific skills needed for peacekeeping, namely, dealing with and engaging local women, then the tendency is to appoint women in those positions. However, this should also be recognized as a possible risk for Afghan women; when they provide information that can increase the general safety in the region, it means that they may also be putting their own kin (sons, cousins, and other male relatives) under scrutiny, which can jeopardize their own individual security. Judging from the number of positive references to MOT Juliette floating around on websites of the SAF and elsewhere, it appears to be a flagship project. However, security problems put an end to the project.

Peacekeeping predominantly involves men because peacekeeping forces reflect the composition of national militaries that are traditionally male dominated. Once within the peacekeeping forces, as in national militaries, men are usually assigned military and policing tasks while women tend toward civilian-like tasks and work as "legal and political advisor, election and human rights monitors, information specialists and administrators. Women's involvement in military peacekeeping remains almost insignificant" (Hudson 2005, 114). According to the Genderforce project, which specifically speaks of the Swedish situation: "Female participation in international service varies depending on the type of mission and ranges between 4.8 percent in the case of Liberia and 11.4 percent in the mission in Kosovo. Many of these women work in nursing and medical care and very few have leadership responsibilities" (Genderforce 2007). In July 2011, that figure was 9 percent women in Swedish international military missions.[9]

The low representation of women in international missions is problematic in many ways. The objectives of UN SCR 1325 calls for an increased participation of women among peacekeeping forces at all levels, but it is equally important that local women are engaged in conflict resolution, peace negotiations, and the peace-building processes. International missions are deployed to help initiate such processes. As I have argued, in cases where local women are subject to oppression and, for example, are allowed to speak only to other women, which are frequent such as in the case in Afghanistan, the mission's success depends on the presence of women in the peacekeeping contingent. If only a few women are deployed in international missions, the objectives cannot possibly be attained, and it matters less whether the peacekeeping forces are considered the most gender friendly. If there are no or too few women to speak with and engage locals, a main objective of UN SCR 1325 cannot be fulfilled. A low number of female peacekeepers also means that the few women present in the organization are, whether they like it or not, needed for and possibly put in charge of tasks relating to the fact that they are women. This means that female peacekeepers are pushed into specific tasks that relate to them being women rather than to their capacity as soldiers or peacekeepers.

CONCLUDING COMMENTS

This chapter verified that Sweden has a gender-aware post-national defense. Gender training has been successful and was attributed to a systematic approach with appointed gender advisors, gender field advisors, and gender focal points and a strong commitment from the leadership. These achievements have been attained without gender parity, as there has not been any significant increase in the number of women in the troops. At the same time, the analysis of the Swedish material fully supports Hilary Charlesworth's (2008) conclusions about the UN peacekeeping context that gender has basically become equated with women. In the context of the Swedish ISAF PRT in Mazar-i-Sharif, gender has been largely associated with the local Afghani women (i.e., gender equals the 'other' women). This way gender is depoliticized into a simple problem-solving tool (Väyrynen 2004, 140; Willett 2010, 144). The problem is solved by including local women in the peacekeeping process, taking their view into account, and making sure that there are a few women in the troops who can talk to the local women. By reducing gender to a simple problem-solving tool, the critical and transformative potential of taking a gender perspective on militaries engaged in peacekeeping practices is lost.

The chapter also pointed to spaces where gender awareness nevertheless had led to or at least opened up the possibility for rethinking masculinity within the peacekeeping forces. This is a positive development. The conclusions are that it is absolutely crucial to understand and discuss conceptualizations of masculinity as well as sexuality to reach the objectives of postnational militaries when they are stationed abroad. Many of the problems related to peacekeeping missions and their legitimacy have little to do with women but rather with masculinity and particularly the problematic view of sexuality embedded in militaries and peacekeeping. It seems important to include as part of gender training constructions of masculinities in relation to femininities and different masculinities in relation to one another (e.g., the Swedish-Afghani masculinities). The way that sexuality, and particularly male heterosexual sexuality, is constructed and understood within the UN, the peacekeeping forces, and the militaries that provide troops for peacekeeping needs to be included in gender-awareness training, as it is intimately related to the legitimacy of the peacekeeping endeavor and key to winning the hearts and minds of the people. To inform of codes of conduct is clearly not sufficient. To equate gender with women has apparently not succeeded in attracting more women to the peacekeeping forces, nor has the appeal to their feminine skills. An understanding of gender that includes and investigates the masculinity, femininities, and sexuality constructions embedded in the institutions of the postnational defense might in the long run be more successful in attracting women to the job of peacekeeping.

CHAPTER 5

Defense and Military Governance in the European Union

In the European Union (EU), there are several indications of a postnational defense: EU's Security and Defense Policy (ESDP), institutions such as the EU Military Committee (EUMC) and the European Defence Agency. A multinational security organization has developed and is based on the contribution of 27 member states. Through the ESDP, the member states assume defense responsibilities for each other, based on the solidarity principle of the Lisbon Treaty, and respond to security challenges in distant locations, far from EU borders. They do this, for example, with help of the EU Battlegroups, which are expected to engage in activities such as peace enforcement, crisis management, and peace building. In 2002, Ian Manners wrote an article in which he considered the EU a civilian and a normative power in international affairs. It may be expected that this civilian and normative stance would influence the emergence of ESDP and EU military activities. Thus, EU's defense and military governance could be considered a cosmopolitan military in this respect, but does it also fulfill other cosmopolitan criteria? Does it include democratic values, such as gender norms, as an integral part of its organizational practices? That is what would be expected if it were to serve as an example to the world and be "a force for good," as Elliott and Cheeseman (2004) put it.

Cooperation on defense issues in the EU offered an opportunity to create something different—a new type of defense that was less militarized and not burdened by the persistent link between national militaries, hegemonic masculinity, and defense practices. Moreover, the ESDP emerged in a context in which gender equality policies and gender mainstreaming were already well known in other policy areas within the EU. This chapter pays close attention to whether gender awareness and gender policies[1] have formed and become relevant to the institutionalization of the military dimension of the ESDP and, if so, how this occurred. To accomplish this, the chapter explores the emerging security and defense governance in the EU by asking how gender has been conceptualized and organized in the development and institutionalization of ESDP and EU Battlegroups.

After a brief introduction to the development of the ESDP, the chapter probes the question of how gender equity has been taken into account in the formation of these new EU institutions. It considers the battlegroup concept and tries to make sense of it from a feminist perspective by investigating images and discourses on gender and militarism in the Nordic Battlegroup—a case within the case of the ESDP. It then looks at how gender has been perceived and discussed in texts on the ESDP. The chapter ends with a discussion of the gender governance in place in the EU and explores whether gender mainstreaming has been incorporated in the institutions of the ESDP and, if so, how.

THE EMERGENCE OF EU SECURITY AND DEFENSE COOPERATION

Although common security and defense concerns have been on EU's agenda for a long time, member states were fairly reluctant to pursue this cooperation—until the failure and inaction of Europe in the Balkans during the 1990s changed this. A collective frustration of sorts initiated the path toward a security and defense policy (Kerttunen et al. 2005, 9). According to Andersson (2006, 13) the weakness displayed by member states during the conflict in Kosovo in 1999 reinforced a notion that if Europe were to play any role in international conflict management, it would require the creation of a common defense policy. This was made possible by the collaboration between Britain and France, who issued a joint statement in 1998:

> The European Union must have the capacity for autonomous action, backed up by credible military forces, the means to decide to use them and a readiness to do so, in order to respond to international crises . . . in strengthening the solidarity between the member states of the European Union, in order that Europe can make its voice heard in world affairs. (Andersson 2006, 13)

Subsequently, when the European Council met in Helsinki in 1999, it decided that the member states must be able to deploy military forces capable of carrying out a full range of tasks, from crisis management to combat. At the same meeting, it was decided that new political and military bodies would have to be established to ensure political guidance and strategic direction for EU operations. The ESDP is thus divided into three components: military crisis management, civilian crisis management, and conflict prevention. To ensure the smooth operation of military crisis management, the Political and Security Committee (PSC) of the council is assisted by the EUMC and the Military Staff (EUMS). The task of the committee for civilian aspects of crisis management is to coordinate and make more effective the various civilian means and resources, in parallel with those of the military, that are at the disposal of the EU and the member states. Finally, conflict prevention aims to identify and combat causes of conflict and improve the capacity to react to nascent conflicts. The analysis here is limited to the military dimension; specifically, the Battlegroup concept, which was initiated in 2003, is one of the various activities conducted within the ESDP and is the focus here.

GENDER PARITY IN THE ESDP: WHY SO DIFFICULT TO 'ADD WOMEN AND STIR'?

Regarding the issue of gender parity, a certain question arises: Does the balance between the number of men and women in the institutions set up by the ESDP reflect the European Council Treaty (Article 2) on equal opportunities for men and women? No, it clearly does not. In the institutional landscape of the ESDP, there is an almost complete dominance of men. Given this, the appointment of Catherine Ashton as the High Representative of the Union for Foreign Affairs and Security Policy is an enormous feat, but she is certainly the odd one. The PSC is the standing political and military committee responsible for the Union's autonomous and operational defense policy, which has a prominent role in the ESDP, and is the most important decision-making body headed by the member states' foreign ministers (Dijkstra 2008). Other institutions are also highly relevant to the agenda-setting and implementation phase of ESDP. For example, EU's special representatives are stationed in different troubled regions in the world. Past representatives have all been men. Rosalind Marsden is the first woman, appointed in 2010.[2]

In addition, the EUMC is the highest military body within the European Council and is composed of the chiefs of defense (high commanders) or their military representatives, and every one of them is a man.[3] The EUMC is responsible for providing the PSC with military advice and recommendations on all military matters within the EU and directs military activities within the EU framework. The chairman of the EUMC attends meetings of the Council when decisions with defense implications are to be taken. Since 2001, the EUMS is a General Directorate within the Council General Secretariat and the only permanent, integrated military structure of the EU. It receives its tasks from the EUMC and normally provides military expertise and support and conducts EU-led military crisis management operations.

Cynthia Enloe (2004, 154) writes about militarism that it "legitimizes masculinized men as protectors, as actors, and rational strategists." This means that men and often militarized men are the ones put in charge of the policies and operations of defense and security activities. This is also what we have noted about the ESDP. Despite the fact that the EU does not have its own military historic past—it is a civilian power—what we see is a direct reflection of the member states' military organizations. Equal opportunities and gender equity do not seem to have been a concern in the set up and operations of these new institutions. Considering the member states provides an explanation.

The armed forces of member states train and provide the staff for international and EU missions and also for the policymaking bodies. Since equal opportunities where not considered in developing the ESDP, the gender gap of the member states' own armed forces is simply and uncritically carried over to the new institutional setup. On the number of women employed in the armed forces, only incomplete data is available in Eurostat figures from 2007. Combining this data with another survey (Council of the European Union 2008d; French Presidency 2008) and data from NATO ("Percentages of Female Soldiers in NATO Countries' Armed Forces" 2007) gives an indication of the extent of the gender gap. The member states' own militaries are almost completely dominated by men—in Finland and Poland by 99 percent. Another 10 member states report that women make up less than 10 percent of their armed forces staff. The largest share of women is in the Hungarian Armed Forces, with 20 percent, and Latvia, with 17 percent; thus, these countries have "only" an 80 percent and 83 percent male dominance rate.[4]

Historically, European militaries have been completely dominated by men. What was argued about the Swedish national military earlier appears relevant throughout the member states. There are different expectations of the citizenship of men and women in relation to military activities. Most often, women have not been required or are not allowed to do military service, while men have been mass mobilized through conscription systems. European militaries were radically transformed after the Cold War

(Joenniemi 2006a). One important change has been the elimination of military conscription, as has been the case in, for example, France, Italy, Sweden, and the Netherlands. Nine member states still have a conscription system in place, although in some, such as Denmark, it is not necessarily put into force. Military conscription duty actually violates equal rights principles. NATO statistics show that it is not possible for women to become generals or serve on submarines in some member states because there are restrictions on women partaking in certain military practices. Some member states, such as France and Britain, exclude women from participation in combat positions. This has significant implications for women's opportunities to pursue a military career and, thus, for the likelihood that they be appointed High Commanders or representatives in the EUMC.

The gender gap is further accentuated in many cases when the personnel of the member states' armed forces are sent on missions abroad (French Presidency 2008). In most member states, the percentage of women participating in international missions is consistently lower than in the forces at home. It ranges from less than 1 percent up to 10 percent. The reason suggested by the Council of the European Union report (2008d) is that women are more reluctant to leave home and go far away. This is not completely accurate. Sweden deviates in this respect from the trends in all other member states, as does Norway. Sweden, with 4.5 percent women in the national armed forces, has on the average 10 percent women in the international missions (French Presidency 2008). Norway shows a similar pattern; it is not a EU member state, but it participates in the Nordic Battlegroup. According to Tore Asmund Stubberud, captain in the Royal Norwegian Navy (February 19, 2010), for the Norwegian Armed Forces, the proportion of women who prefer to go on international missions is as high as 15 percent, compared to those who want to only do military duty at home (7 percent).

However, in the Nordic Battlegroup in 2008 only 5 percent of the troops were female. Other EU missions showed a similar pattern. In the European Military Force (EUFOR), female representation was just under 6 percent; in the EU Police Mission, it was 8 percent (Valenius 2007b, Annex Tables 1 and 2). In the Democratic Republic of Congo (RD Congo) missions of the EUFOR and the EU Common Security and Defence Policy (EUSEC), the female share of the mission staff never exceeded 5 percent (Gya et al. 2009, 14). This gender imbalance of the EU member states' national militaries explains the gender gap in EU missions, battlegroups, and decision-making bodies. The high impact of national military organizations at the EU level leads me to conclude that the ESDP in this respect falls short of being considered a cosmopolitan military. It has basically mirrored the national conditions, and there is no indication that anyone is seriously concerned about it, despite the prominent role of equal opportunities in the European Council Treaty, referred to also in the ESDP documents.

The pursuit of gender equity was not salient on the agenda when the ESDP and related institutions were established. This is indicative of the way a gender order is reproduced. If the gender order is not questioned, the historic connection between militarism and masculinity is carried on automatically through what appears as normal practices (cf. Connell 1995, 212). Indeed, the historic gender gap of member states in regard to their military and defense sectors has been simply carried over to a new European institutional arrangement.

This explanation is also related to the EU decision-making process regarding security issues, which relies on consensus with respect for the sovereignty of member states. Hence, it would be inappropriate to question the way member states organize their military forces. Having said that, it remains perplexing that virtually no attention has been paid to this issue in the development of ESDP policy and institutions. Gendered military practices, in member states' militaries, greatly impact the career opportunities and the possibilities for women to take part in the agenda setting, decision making, and implementation of EU's security and defense politics. It reduces women's agency in the ESDP and puts the influence in the hands of institutions of hegemonic masculinity, the traditional militaries of the member states. Women are deprived of agency because they are not appointed to leadership positions and/or because they do not have complete and equal access to the training and qualifications that are basic requirements for positions in the EUMC or EUMS, for example; nor are they encouraged to join ESDP missions. Although we cannot equate women with having and proposing feminist values, empirical evidence shows that gender mainstreaming rarely happens without women standing up for those ideas.

Catherine Ashton, the first female in a decision-making position within the ESDP, has by her recent nominations of women to leadership positions given evidence that women to a larger extent nominate women and put gender issues on the agenda. Margot Wallström, former vice president of the EU Commission, has criticized the EU for being what she called "a reign of old men," referring both to the proportion of men in decision-making positions in the EU and to the specific way in which recruitment is conducted ("Margot Wallström Fed Up" 2008). She described EU recruitment processes as homosocial relations (i.e., men favoring men). In the context of military relations, this is certainly true. There is a military male presence with agenda-setting power in EU security and defense relations, accompanied by a fairly compact silence on both equal opportunity and gender parity concerns.

In gender studies, it is often assumed that gender parity is easier to achieve than gender mainstreaming because the former is about adding women. It does not necessarily call for a questioning of the values and norms embedded in the institutions themselves. For security and defense governance, gender parity indeed seems very difficult to achieve. The following section goes into more depth on how militarism and gender are embedded in ESDP practices, with a focus on the EU Battlegroup.

WHAT DO BATTLEGROUPS DO FOR A CIVILIAN POWER?

The introduction of battlegroups as a component of the ESDP is puzzling from the perspective of the EU as a civilian power. Operation Artemis in 2003 was the first operation within the ESDP that the EU conducted on its own. It took place in Congo and was considered successful because it showed EU as a postnational defense organization that had the capacity to react quickly and with force (Andersson 2006, 10; Lindstrom 2007, 10–11). The operation Artemis set an example and became important for the development of the battlegroup concept in ESDP. In light of the EU's usually long-winded policy- and decision-making process, the agreement to form EU Battlegroups is quite remarkable. It materialized very quickly: The idea was raised in 2003, by 2004 a majority of the member states had agreed to form battlegroups, and by the beginning of 2005 the EU had a rapid reaction capacity (Andersson 2006, 21).

An EU Battlegroup consists of approximately 1,500 troops with combat support and logistics units as well as air and naval components. The goal is that they be ready for international deployment within five days after approval by the Council. EU member states agreed to have two battlegroups on standby for six months at the time. The battlegroup is a specific form of rapid response in the EU defined as the smallest coherent force that is at the same time both military effective and rapidly deployable, as well as capable of stand-alone operations or at least for the initial phase of larger operations. The battlegroup is based on a combined arms and battalion-sized force and is reinforced with combat support. A battlegroup can be formed by a framework nation or by a multinational coalition of member states. In all cases, interoperability and military effectiveness are key criteria. A battlegroup must be associated with a force headquarters and have preidentified operational and strategic enablers, such as strategic lift and logistics. In most cases, the battlegroups are multinational and expected to take on the whole range of tasks expected from a postnational military, such as humanitarian assistance, peacekeeping, and peacemaking, as well as conducting high-intensity combat operations. Sweden, Finland, Estonia, and Ireland together with the non-EU member Norway established the Nordic Battlegroup, which was on standby for the first part of 2008 and 2011. It was under Swedish leadership and was the first multinational battlegroup composed entirely of smaller EU member states, including a non-EU member. (Andersson 2006; Lindstrom 2007; Kerttunen et al. 2005).

To the laymen the concept of a battlegroup sounds militarist, but it is a well-known military term, and in military language it defines a specific organizational unit. The term seems clear and uncontested, for example, in Andersson (2006, 22) who uses the concept of a battlegroup as a taken-for-granted organizational setup, rather specific in terms of personnel, equipment, and support and thereby rather "naturally" accepted by defense strategists and perhaps defense politicians across Europe. Lindstrom (2007) also suggests that there is some type of basic understanding of the concept as an organizational term, while there are different meanings regarding the size and capabilities of such a battlegroup and variations in how it is viewed, for example, in the air force and the army (Lindstrom 2007, 13) or the navy (Kerttunen 2005, 29). These studies as well as the original document on the EU battlegroups lack reflections on what the choice of concepts might mean for the development of security and defense policies of a 'civilian' power—the EU. Rather it appears that militarism has been instrumental in developing the battlegroup; it is a term derived from a transnational military language and organizational tradition, uncritically adopted and simply taken on in the context of the EU. Ian Manners (2006, 189) in his later work suggests that the EU may be moving away from its ambitions as a civilian power by pointing to this militarization of the EU (see also Ellner 2008). The battlegroup may be an example of this.

Militarism has been discussed in previous chapters. Of special interest here is how militarism contains norms on how actions should be carried out and organized, for example, in battlegroup-size operations, and the idea that this understanding is similar across European state militaries. Similar organizational features among participant nations increase interoperability in multilateral forces and are learned through military alliances and joint operations. The battlegroup concept has emerged out of a traditional military way of thinking. Being part and parcel of the ESDP, it can be argued that militarism has been institutionalized in the EU rather uncritically.

As noted previously, Sweden took on the responsibility and leadership of the Nordic Battlegroup, a multinational EU military force, on standby in 2008 and 2011. Sweden's promotion of the battlegroup concept is an interesting break from the past. It is a contrast to how Sweden perceived itself as a neutral, defensive actor engaged in international relations as a peacekeeper and crisis manager, promoting conflict prevention (Björkdahl 2008, 2002). While the multinational character of the Nordic Battlegroup and the commitment to a common EU defense and security fits the postnational dimension of Sweden's defense (Jacoby and Jones 2008), the association with combat activities in the way of the Nordic Battlegroup is a development that has evaded public discussion.

This has also been heavily criticized, for example, by the Swedish branch of a nongovernment organization, the Women's International League for Peace and Freedom. The League is a feminist, antimilitarist organization, very critical of the militarist values that underlie the concept of and the activities to be carried out by battlegroups. The organization questioned how Sweden, a neutral, defensive, international activist state—which was a stance widely supported by a majority of the public—suddenly without any public debates became a nation ready to go into combat situations in distant places (Moberg 2008). According to Kerttunen et al. (2005, 54), the Swedish and Finnish foreign ministers faced a similar dilemma when the development of EU's defense was on the agenda. The authors argue that in an effort to divert the EU from pursuing a full-fledged, common defense strategy, the two foreign ministers from Sweden and Finland pushed the idea of battlegroups because they viewed it as a limited version of a common military defense. This was more acceptable to the former neutrals, Sweden and Finland. Thus, for them the development of the Nordic Battlegroup was a way to show willingness to support the process of building the ESDP while simultaneously trying to keep militarism at bay by limiting and containing those forces—to make the EU's "military bite" more specific and contained (Engström 2011: 171–197).

Indeed, the EU's military activities through the battlegroups have certainly been contained: to date no battlegroup has actually been deployed on a mission. On the other hand, it can also be argued from a cosmopolitan viewpoint that if the EU will not take on the responsibility of protecting distant people in the name of human rights, then the risk is that no one will or can. In the following sections, the Nordic Battlegroup is used to illustrate how militarism and gendered practices can become embedded within what we have argued here is a new institutional development within the EU.

An episode that evolved around the selection of the insignia for the Nordic Battlegroup illustrates the importance of military traditions in contemporary military practice and shows how those traditions gender the battlegroup in the symbolic representation of its identity. Heraldry is part of militarism and military culture (van Creveld 2008). Heraldry regulates how coats of arms and armorial bearings are devised and described, that is, it relates to the use (and affection for) emblems, flags, symbols, and ceremonies. Such traditional practices turned out to have importance in the Nordic Battlegroup, a completely new military constellation.

In an interview, Nordic Battlegroup Force Commander Karl Engelbrektson (2008) related the coat-of-arms story, which is paraphrased here:

Long before the Nordic Battlegroup was actually set up, its future leadership felt the necessity to develop an insignia. It would be important to unite and create a common identity across the multinational troops with a unique symbol—a coat of arms. The insignia should be one that could unite four countries and the colors would reflect the countries involved. It is important to turn to heraldry on occasions like this, Engelbrektson explained, because there are sets of symbols with specific meanings. The heraldic lion was chosen because it represents strength. It held a sword, to signify that the battlegroup was prepared to use violence if necessary, as well as an olive branch, to signal the priority to use peaceful means. Swedish heraldry has strict rules; the lion must be a male with a large mane and a visible phallus. The Force Commander said that, even if they wanted to use a lion for strength it was *not* necessary to exaggerate this with a phallus. To not include one would be a way to show sensitivity to UN SCR 1325 thinking.

Another version of the same story appeared in the SAF's newsletter (*Insats & Försvar* 2007:5, 8), and it goes as follows: "According to military tradition, if a lion is to be part of an insignia it must be a male lion with a phallus. This was something that offended the female staff in the Nordic Battlegroup, and they reacted to it. They wanted a gender-neutral lion in the insignia. The traditionalist men protested against this violation of a long military tradition. However, Javier Solana supported the wishes of the female staff and the lion was 'castrated'" (my translation).

The narrative affirms that the battlegroup, although a new institutional creation, incorporates and grapples with military norms that permeate military institutions and practices in the member states. This version of the story circulated in the international press, and heraldic experts entered into the debate. Another article (*Göteborgsposten* 2008) tells us that to 'castrate' a heraldic lion like this is a severe violation. Historically, lions without visible genitalia were given to those who violated the Crown; hence 'castration' is a highly disgraceful act.

There is much to learn from these stories. For one, it is clear that symbols, insignias, and military heraldics are important in the battlegroup, for example, to create a sense of collective identity. It is part of militarism, and it is gendered and sexed. The connection in heraldics between strength and the male phallus is important, as to increase what is clearly a male lion's potency is to make sure the phallus is visible. The mane is not a powerful enough expression of strength. The importance of the phallus in the military is also highlighted; to be 'castrated' is to be emasculated, and that is dishonorable and abhorred. Military potency thus lies in an intact (heterosexual) masculinity.

Different versions of the story circulated. According to Engelbrektson, the twist to the story that emphasized that the lion was castrated based on women's demands (*Insats & Försvar* 2007:5, 8; see also "Army Castrates Heraldic Lion" 2007) was fabricated. He said there were no women involved in the decision on the insignia, as no staff had yet been recruited. This suggests that different factions exist within the military. The military traditionalists, as they are called in the article, seem to fear a feminization of the Nordic Battlegroup. That the lion was said to have been 'castrated' by the women with the support of EU may have been a way for traditional military men to protest against the inclusion of women, gender perspectives, and UN SCR 1325. They protested the emasculation of the military. The traditionalists were perhaps also disturbed by the move toward a postnational military. The SAF article cited above, referred to Javier Solana, at the time the prime expression of EU authority, who supported the women's wishes. The traditionalists were protesting how authority over the military was being denationalized through the EU via the battlegroup. The different versions of the insignia narrative suggest there is not full-fledged support of the developments toward a postnational military. On the contrary, the traditional military men fear 'castration' and perhaps mourn the traditional Swedish defense: 'neutral' but certainly all male and masculine.

The image (Figure 5.1) showing a man overlooking two women sewing the coat of arms onto the Nordic Battlegroup flag seems to depict that traditional man who, concerned with upholding proper military traditions, watches to be sure it is properly done: no 'castration' here. Hence, militarism lives on also in the new institution, the Nordic Battlegroup. However, it is clear that there are divisions within the organization. There is a tension between traditionalists supporting the national military as it was and those who aim for a more cosmopolitan-minded approach to contemporary military practice and institution building.

Figure 5.1: Swedish Armed Forces, Photo Torbjörn F. Gustafsson

This photo is interesting in many respects. When I first saw it, I immediately associated it with the photos that were frequent in the documentation (newsletters, pamphlets, and information booklets) that I had come across when researching the voluntary defense organizations. Indeed, it evoked a sense of the neutral national military during the Cold War and the gendered tasks associated with it. How could this be part of the newly launched postnational Nordic Battlegroup?

THE NORDIC BATTLEGROUP: THE DOMINANCE OF MASCULINITY AND MEN

I conducted a closer analysis of the Nordic Battlegroup by studying a special issue on it from March 2007 in the SAF newsletter (*Insats & Försvar* 2007:3) and an information pamphlet from the SAF on the Nordic Battlegroup. The material is packed with drawings and photographs that made it possible to consider how women and men were represented in the material. Most photos depicted men; there were in total five photos of women. There were no women included in any group or military action photos; none depicted women in situations as active soldiers. There was no evidence of a specific women-in-arms identity articulated for the battlegroup. The only action-oriented photo showed the upper body of a female officer in combat fatigues added to a collage of photos illustrating a training event. Two women were portrayed in a set of portrait-style photos related to interviews with different soldiers of the Nordic Battlegroup. On the other hand, the two remaining photos of women depicted them doing traditional feminine tasks. In one photo, a woman is assisting at an operating table, and in the other, two women are sewing the Nordic Battlegroup insignia onto a flag (Figure 5.1 above).

In the efforts of the SAF newsletter to explain and describe the Nordic Battlegroup, a norm of the Nordic Battlegroup soldier emerges. It appeals to a form of combat masculinity that for a long time has been part of the military culture (Higate 2002). Chapter 2 discussed how soldiers require intense socialization and training to fight effectively and that gender identity becomes a tool that societies use to induce men to fight (Goldstein 2001, 252). Already the recruitment campaign for the Nordic Battlegroup had used a militarist and gendered imagery in advertisement and presentations. The emphasis on a certain kind of 'warrior-like' masculinity in the recruitment campaign for the Nordic Battlegroup may be related to this fact: To be successful in recruitment, the classic expectations on men as the soldiers is evoked, and the construction of masculinity is used in advertisement[5] as a way to attract a specific kind of person. The image of the Nordic Battlegroup soldier conveyed a masculinity that evoked a warrior rather than a cosmopolitan-minded peacekeeper, which also became normative for the Nordic Battlegroup 2008 (*Insats & Försvar* 2007:3).

As is evident in the material presented earlier in the text, battlegroups are expected to take on a wide range of tasks: humanitarian assistance, peacekeeping, and peace making, while also being prepared to conduct high-intensity combat operations. To focus too much on combat is problematic. As stated in Liora Sion's (2006, 471) study of Dutch peacekeeping units, the military organization tends to train for military combat activities even if the requirements in international missions are different. Force Commander Karl Engelbrektson found this a major challenge too: "In the military training at home, we teach soldiers to use as much weapon power as possible to get the best effect as quickly as possible, while in international missions the task is the opposite: to calm aggression and thus, to shoot as little as possible." (*Insats & Försvar* 2007:5; my translation). Since combat training is highly masculinized it reinforces the connection between masculine identity and military activities. This seems to be the case also in the EU context through the battlegroups. Furthermore, the way that the Nordic Battlegroup was presented to the general public and prospective recruits was likely to discourage rather than encourage women to sign up. Neither the newsletter nor the recruitment adverts presented a 'woman-in-arms" image for women to identify with. As it turns out, few women were interested. Although the goal was set at 20 percent, only 5 percent of the recruits for the Nordic Battlegroup were women, less than the average of other international missions. The threat of 'feminization' as articulated by the traditionalists seems far-fetched, and the ambition to recruit women into the Nordic Battlegroup not genuine.

THE CONCEPTUALIZATION OF GENDER IN EU'S SECURITY AND DEFENSE POLICY

Official texts of the ESDP provide a general idea of how gender was perceived as the policy took shape and institutions developed. I conducted a content analysis of a broad range of documents relating to the EU's decisions and actions in the field of security and defense published in two collections (EU Security and Defence 2005, 2007). I also studied documents relating to EU's strategy on UN SCR 1325 and 1820 (Council of the European Union 2008a) as well as information made available on the Council of the European Union's website: Women, Peace and Security.[6]

The result of the analysis of the official documents in the two collections was that, for the first period (EU Security and Defence 2005), gender was mentioned four times in 420 pages of documents. Three times the word *gender* was mentioned simply as one among other concerns in a list of desirable elements to be considered in peacekeeping and crisis management operations. On one occasion a statement mentioned more concretely that gender perspectives should be a part of training in ESDP. Gender was not discussed or explained as a concept. On the other hand, the word *woman* appeared 10 times. In 8 of these cases it regarded the situation for women as victims of violence in countries outside Europe with ongoing conflicts and where peacekeeping operations may be needed. Only twice did the word *woman* come up in reference to values important to the EU (i.e., equality between men and women, as in Article 2 of the Lisbon Treaty) and those were also the only times the word *men* appeared in the text (EU Security and Defence 2005). The analysis suggests that, in 2004, the EU policymakers involved in ESDP chose not to problematize or were unaware of the gendered aspect of militarism, defense, and security. Gender remained on the list of desirable goals. Women were, when mentioned, considered victims of violence in faraway places. While equality between men and women at the time was an important goal for the EU, it remained effectively outside the EU defense and security field.

The same type of analysis was done on the documents relating to EU's decisions and actions in the field of security and defense for the next two-year period (EU Security and Defense 2007). Some changes were noted. Gender was mentioned far more often—41 times in the 466 pages of collected documents. The increase can be attributed to a specific document that detailed the implementation of UN SCR 1325 and gender mainstreaming in the Presidency Report on ESDP. It outlined how the EU intends to implement resolution UN SCR 1325, although it was not until 2008 that the EU issued its own action plan for the implementation of UN SCR 1325 and 1820 (Council of the European Union 2008a). There is a connection between the work with UN SCR 1325 and gender mainstreaming in these documents (EU Security and Defense 2007). There were 28 references to women in the document, and, as with the earlier documents, women were discussed as victims. The references were related to women's situations in non-European countries as specifically vulnerable and often in connection with children. Again, the problem with women and gender was related mainly to aspects outside the EU. On the one hand, this signals a postnational approach with the view 'to save distant others', but it is a view that implicitly associates gender and UN SCR 1325 only with vulnerable women beyond EU's borders.

On the other hand, the word *men* appeared seldom in the documents—only five times: four times in connection with women as in "men and women" and once as a substitute for "personnel" in "men and resources are lacking'." Men and masculinity are two concepts that are virtually silenced and, as argued previously, this silence is an indication that masculinity and men are perceived as the normality of defense and security governance. One characteristic of gendered militarism is that the category of "men" is never questioned or politicized. Men and masculinity are simply included in the norm system of the ESDP and, thus, need not be written about; the routine maintenance of the institutions will continue to reproduce the norms (Kronsell 2006, 2005). Women, femininity and gender, on the other hand, are treated as something external. In these documents, women are discussed, if at all, in terms of outsiders. Norms of masculinity seem to prevail as codes of governance of military, defense and security, and men are normalized as the given agents of the ESDP. Yet a clear trend is that the notion of gender is gaining acceptance. This is mainly in relation to UN SCR 1325 and to peacekeeping operations outside the EU.

There is a focus on gender mainstreaming even though it is limited to certain sections of the documents. It appears as if gender mainstreaming strategies and work started on a more serious scale in 2006. Having said that, it is clear that the ESDP had not fully incorporated lessons learned from gender and feminist analyses of militaries, despite the fact that such research was widely available at the time. The quote from the report on the Implementation of the European Security Strategy (Council of the European Union 2008c, 10) exemplifies the EU's position:

> We need to continue mainstreaming human rights issues in all activities in this field, including ESDP missions, through a people-based approach coherent with the concept of human security. The EU has recognised the role of women in building peace. Effective implementation of UN SCR 1325 on Women, Peace, and Security and UNSCR 1612 on Children and Armed Conflict is essential in this context.

Women are perceived as outside the ESDP and are not viewed as political agents, but they need to be included in peace-building processes where there are armed conflicts (i.e., outside the EU). Mainstreaming has relevance but it is not about gender, it has turned into mainstreaming of human rights (see also Council of the European Union 2008b). Women are still relevant in peace building and in the practice of ESDP on location in 'distant places' but *gender* entirely disappears from the agenda and is replaced by a people or human security approach.

GENDER GOVERNANCE IN THE EU

As argued initially, many different strategies have been developed to include women and gender in policymaking. Gender mainstreaming, equality norms, and equal opportunities are the means used to further EU gender governance —so far with varying success. Gender parity is often a goal and also one adhered to in political parties and institutions (Dahlerup and Freidenvall 2008). For democratic purposes and for the purpose of securing women's presence and power in decision making, it is important (Phillips 2000; Young 2000). Gender parity in terms of a balanced representation of women and men in all areas of EU activities are part of the fundamental rights and principles of the EU, laid down in Article 2 of the European Council Treaty. The European database on the representation of women in decision making pays witness to the gender gap and imbalance in women and men's representation in politics but also in a different decision-making context across Europe.[7] Of this, the Swedish Presidency report (2009, 94) writes:

> While the last few years have seen a general increase in the number of women in decision-making positions in Europe, women continue to be a minority in the

political and economical spheres. In parliaments, governments and ministries, and in the private sector too, power still rests in the hands of men.

As has been pointed out previously, there is an almost complete lack of gender parity in EU defense and security governance.

Another area of gender governance is equality legislation. Equality norms have, through European integration, multilevel governance and with the use of the court system been rather successful in pushing equality law forward, at least in the economic and trade sector. Anna van der Vleuten's (2007) work shows how equality norms were developed and Europeanized. Reluctant member states were pinched and sandwiched between the various legal and political levels of EU policymaking—for example, pushed by national and transnational women's groups and the European court simultaneously. However, decisions taken in the field of security, defense, and military relations are intergovernmental, built on consensus and the will of the member states. They are not subject to the rule of the Court or decisions in other areas, although it may nevertheless be assumed that all policies within the EU cooperation would take gender issues into account for democratic and legitimacy reasons. What is new with the Lisbon Treaty of 2009 is that gender mainstreaming also applies to the ESDP.

Gender mainstreaming is yet another strategy for gender governance. EU's Fourth Action Programme on Equal Opportunities for Women and Men (1996–2000) "featured mainstreaming as the single most important element" (Hafner-Burton and Pollack 2002, 295). The underlying thought of gender mainstreaming is that most activities have a gender dimension and that gender relations relate to underlying norms of political and administrative institutions (Verloo 2001; Walby 2005; Woodward 2004). Gender mainstreaming calls for gender perspectives to be incorporated into all policy areas, stages, and levels and emphasizes the need to look at gender, not just women. Gender mainstreaming has a radical potential because it widens the policy frame and expands ideas about "the broader structural and institutional causes of gender inequality and discrimination" (Liebert 2002, 250). As a governance principle, it differs much from equality norms because it has this radicalizing potential but is also a policy strategy and a soft law instrument (Jacquot 2010; Mazey 2002). In the EU, *soft law* refers to the various policies that are not binding on the member states but nevertheless indicate a policy direction or intention. Eveline and Bacchi (2005, 506) suggest that gender mainstreaming should be envisioned as a process of gendering that is always incomplete and something that "people-as-bodies 'do' through their practices" and in relation to others. In this way, gendering is seen as an embodied process that relates simultaneously to male/female bodies and the construction of femininity and masculinity. The idea is that gender mainstreaming can transform the entire organization. This is why we should expect it to have relevance when new policy areas emerge in the EU, such as the ESDP.

The success of gender mainstreaming in the EU context, Maria Stratigaki (2005, 178, 181) argues, has been hampered precisely because it requires fundamental changes in long-embedded social relations and challenges the gender distribution of power. Gender mainstreaming is a strategy that has originated from and is a part of transnational women's and feminist movements (cf. Prügl 2009; Rees 2005; Walby 2005), and since the concept has become embedded in many states' and international institutions' policymaking, the efforts of the transnational women's movement to push for gender mainstreaming has succeeded (Woodward 2008, 289; Zalewski 2010, 5).

As pointed out earlier, UN SCR 1325 and 1820 have in the EU context been connected with gender mainstreaming. The strategies to implement UN resolutions in the ESDP are specifically related to international missions. The EU's own Action Plan on UN SCR 1325 is the comprehensive approach on women, peace, and security that was adopted by the EU Ministers for Foreign Affairs in December 2008 (Council of the European Union 2008a), eight years after the resolution was adopted in the UN. As we saw in the previous analysis of the ESDP documents, discussions on UN SCR 1325 came earlier and were captured in the texts beginning in 2005. Parallel to this, 12 of the 27 member states have developed national action plans on UN SCR 1325 and 1820 that apply to their own domestic activities in security and defense issues. This includes Sweden. That these strategies have been developed can, as is the case with gender mainstreaming, be considered a triumph for women's and feminists' transnational political activism.

In the Swedish context, gender mainstreaming has been a part of public policies since the governmental bill in 1994 (Proposition 1993/94:147). It requires that public authorities, including the defense forces, gender mainstream their policies and strategies. However, it has been difficult for many Swedish authorities to follow the legislation on gender mainstreaming. In a recent evaluation of gender mainstreaming activities in public authorities, 69 (n = 190) reported that they so far had done nothing (Nationella sekretariatet 2009, 26). That is, 14 years after the legislation had come into effect, over one-third of the authorities still had not implemented gender mainstreaming: this, despite that there has been wide political support for gender equity issues in the Swedish context for some time (Sainsbury and Bergqvist 2009). Röhr et al. (2008, 18) write of gender mainstreaming that it is not an accident that this strategy often is resisted because, if put into effect, it is "where the most efficient structural change to the benefit of gender equality will be found." The potential in the approach lies in the fact that "the burden of proof is reversed." Gender mainstreaming requires that institutions, through their own means, complete the tasks necessary to demonstrate that they are gender neutral. The goal of gender mainstreaming is to liberate the institution of gender biases, and this threatens male privilege and power.

GENDER MAINSTREAMING THE ESDP THROUGH TRAINING

A report by the Council via the French Presidency (2008) suggested that a majority of the member states were engaged in some gender mainstreaming activities in the area of defense and security. Data provided by member states as an input to the Swedish Presidency's (2009) report to Beijing +15 helped qualify these findings. A major part of the actions taken and reported on gender mainstreaming in relation to the military, defense, and peacekeeping were in the response to and in the context of UN SCR 1325. The member states did not explicitly connect this with a general gender governance strategy for the EU. Although there are similarities, the work on gender mainstreaming and training in UN SCR 1325 diverges. With UN SCR 1325, the emphasis is more on how to include, interact, and relate to 'vulnerable' women in conflict and war-torn areas outside the EU and deals much less or not at all with gender as part of the organizational activity of the military and ESDP institutions.

Johanna Valenius (2007b) studied gender mainstreaming in ESDP missions in Bosnia-Herzegovina and concluded that gender mainstreaming is not well understood by policymakers and practitioners in the ESDP. There is a significant gap between policy and practice. The EU Police Mission and the EUFOR Althea in Bosnia-Herzegovina experienced difficulties in going from rhetoric to practice. Senior military officials tended to be positive about women's participation on a general level, but practices on the ground revealed a lack of understanding of gender issues. Stereotypical ideas about women and men were frequent and influenced the views of what women would add to the military or what problems they would create. Despite the rhetoric, the tendency in practice was to be skeptical about women's participation and contribution to missions (Valenius 2007b, 33–38).

Similar thoughts were conveyed at a 'taking stock' EU conference in 2009 on UN SCR 1325 and 1820 training in ESDP missions and operations. Based on their experiences in missions, the participants particularly emphasized the need to, through training, go from gender as an intellectual idea to understanding gender as something relevant in practice in the everyday activities of the operation. The view conveyed here also was that gender as an intellectual idea had little relevance for ESDP missions. If gender can be framed as a win-win solution in that it becomes a way to increase operational effectiveness, it can become part of the everyday practices of the missions (Folke Bernadotte Academy 2009).

The conference took its starting point in a study based on surveys on the training in UN SCR 1325 and 1820 in the member states and in a number of missions. A finding was that "much of the current training appears to be on the level of awareness raising and less on the practical application of the resolutions" (Folke Bernadotte Academy 2009, 20). From the surveys, the authors noted that the training on UN SCR 1325 and 1820 was usually only focused on discussing the importance of equal representation and opportunities of women in the missions and how to behave toward local women as part of cultural awareness training. Other important issues in UN SCR 1325 like: ensuring local women's empowerment, women's participation in peace processes, and women's peace work was rarely part of training, according to Folke Bernadotte Academy (2009, 21) issues dealing with sexuality, masculinity, and violence were not either. Another interesting report on how UN SCR 1325, 1820, and gender perspectives were implemented examined ESDP missions in the Democratic Republic of Congo (Gya et al. 2009) and accounted for a number of best practices and success stories, particularly highlighting the relevance of gender advisors. The authors point to the necessity of having gender advisors and gender field advisors in the missions who can train people and point out gender relevance in daily operations of the missions. Currently, almost all civilian and military ESDP missions have one or more gender advisors. Yet the general findings in the Gya et al. (2009) report is that while there are gender advisors and gender field advisors, so few resources are dedicated to these functions that it is difficult for them to do their job in a satisfactory way. Furthermore, they report that the call to gender mainstream is hardly taken seriously. UN SCR 1325 and 1820 are perceived as optional policies, considered side issues, and seldom part of the general planning or approach of the missions.

In preparation for the first Nordic Battlegroup, the Swedish Defense College formulated some recommendations. Ivarsson and Edmark (2005, 135) made a number of recommendations to the Nordic Battlegroup 2008 on how to pursue the ambitions of UN SCR 1325. These were taken to mean three things for the Nordic Battlegroup: first, that there should be an increased recruitment of women, particularly in leading positions; second, that gender issues should be a part of the military education and training of the troops; and third, that code of conduct should be applicable to the activities of the troops on a mission (Ivarsson and Edmark 2005, 129). The EU has such a standard for all ESDP operations, and the Swedish government proposed a code of conduct applicable to all Swedish citizens on international missions. This code of conduct, discussed in the previous chapter, applied to the entire Nordic Battlegroup because Sweden was heading it. Because the battlegroup was not deployed, the code of conduct was never put to the test.

A general observation can be made: Although the strategies outlined above fit with what would be considered gender governance and gender mainstreaming, it is important to note that there seems to be a preference in practice in Sweden to avoid referring to terms such as *gender mainstreaming* in relation to defense and security activities that are to take place in the international context; often referred to instead are UN SCR 1325 or simply 1325. For Sweden, there is a long tradition of gender mainstreaming in politics and administration, a tradition that is clearly associated with feminist politics and femocrats (Sainsbury and Bergqvist 2009). It may be that the SAF prefers not to associate with this more political concept. Referring simply to "1325" makes it possible to engage in a limited kind of gender mainstreaming activity that is more specifically related to the defense and military activities and can be more narrowly defined as a gender perspective that is about the treatment of women in faraway places and their protection from perhaps violent and patriarchal men from other cultures.

The first recommendation that Ivarsson and Edmark (2005) listed for the Nordic Battlegroup—to increase the recruitment of women—was not fulfilled. As already mentioned, out of those who served in the Nordic Battlegroup in 2008, only 5 percent were women, and this is much lower than the 20 percent aspired to. Indeed, it was difficult to improve gender parity. However, two women were appointed to leadership positions; one was a gender advisor directly responsible to the commanders. Both women lacked a military background.

Enactment of the second goal—to train the troops in gender issues and particularly in connection with UN SCR 1325—was successful (Engelbrektson 2008; Jufors 2008). There was extensive training at all levels of the Nordic Battlegroup organization. It was up to the appointed gender advisor to both advise and monitor the work on gender mainstreaming. She was directly responsible to the force commander, and this made her position particularly influential. She was very satisfied with the extent of the educational efforts throughout the entire Nordic Battlegroup organization. On the SAF webpages in early 2008 you could read the following about her:

The Gender Advisor, Madeleine Jufors, was a county police commissioner formerly responsible for the police in the region of Blekinge. She will watch the situation and advise the Force Commander concerning gender matters. In an interview she said: "In short my job can be summed up as reflecting the importance of having women in the unit both to be able to reach women in the area of operations and to keep a watch on the situation of local women and children, since in a war zone they often have to manage on their own and are at risk of rape and other abuse" (my translation).

That the gender advisor held the rank of lieutenant colonel and was responsible only and directly to the force commander gave her the necessary status to pursue work with UN SCR 1325. It was also very important that the leadership and the force commander were convinced of the importance of this. This in turn is related to the work conducted in the Genderforce (2007) project, which has been very important to the development of gender advisors. This project, now finished, has been conducted in collaboration with the SAF, the Police Force, the Rescue Services Agency, the association of military officers, the women's voluntary defense organization, and the Kvinna-till-kvinna foundation. The purpose of the project was to improve Swedish international operations and missions in areas of crisis and conflict by increasing women's involvement in peace support and reconstruction processes as well as in the actual operations and missions.[8] It was about making gender knowledge operational. The project has provided educational material on the website and has trained gender coaches and advisers who are expected to conduct the work on gender mainstreaming in the EU and other missions. The work built on the experiences that Sweden had gained previously in peacekeeping missions in Kosovo and Afghanistan.

Particularly interesting is the 2006 project on gender coaching that was geared toward the top level of the SAF. Six generals and the director of military education were assigned a gender coach with whom he or she could discuss practical problems and exchange ideas (*Insats & Försvar* 2006:6, 58). The fact that gender coaches and gender advisors work with top management is important as it gives legitimacy to gender mainstreaming inward in the organization and outward to people around the missions. The appointment of gender advisors, as we see in the Nordic Battlegroup, was geared in a similar way to integrate gender awareness at the top level of the organization.

In the previous chapter, the Swedish Provincial Reconstruction Team of the International Security Assistance Force was referred to as a good example of a gender-aware cosmopolitan military. Part of its staff had previously been in the Nordic Battlegroup. This can be taken as a positive evaluation of the Nordic Battlegroups' progress in training. The education and training conducted there seemed to have more permanent effects on the leadership and the recruits' knowledge of gender issues and also on the way that gender and UN SCR 1325 training was set up in the organization (Olsson and Tejpar 2009).

A report that analyzed the outcome of gender mainstreaming in state authorities in Sweden explored the reasons behind successful gender mainstreaming. A major portion of the respondents who had experienced successful mainstreaming pointed to the role of leadership and to the importance that top management supports gender mainstreaming (Nationella sekretariatet 2009, 30). This is also what was argued about gender mainstreaming in the context of the battlegroups. The leadership effect may be particularly forceful in the context of the military, an organization that is known to be strictly hierarchical in its structure. Kathleen Iannello (1992, 15) writes that hierarchy describes a top-down structure with the delegation of power as well as a determined set of functions. The hierarchical organization of militaries has hardly been questioned. It is often argued that the demands of war makes the ability to communicate messages swiftly in the organization and the need for each member to be sure what functions he or she needs to fulfill makes hierarchy absolutely crucial for military performance. However, feminists and antimilitarists have criticized hierarchy as an organizational form.

Cynthia Enloe (2007, 4) considers the adherence to hierarchy in military governance as closely tied to militarism. Her criticism is that human needs rarely are part of the hierarchical organization, nor are relations between members of the organizations outside the hierarchical authority structure fostered or nurtured (Iannello 1992, 21; cf. Ferguson 1984, 22). Thus, hierarchies lack flexibility and could have difficulties in addressing problems related to personal issues, the civilian sector, or the local population. Yet one benefit that hierarchies have and the respect for authority chains lead to, as we have seen here, is that if gender issues are accepted and promoted by the leadership at the top, the likelihood that gender mainstreaming will permeate training and the activities of the organization increases. In the Nordic Battlegroup, the example was that the gender advisor was directly linked to leadership and also very supportive of gender issues and thus was fairly successful in influencing the entire organization.

It is often argued that gender balancing is a simpler strategy (i.e., more likely to be implemented) because it fits with liberal political ideas of democratic governance and equal representation. Gender mainstreaming, on the other hand, is viewed as a more radical strategy because it calls for an analysis of all types of policies and decisions and has as its basis an understanding of gender as socially constructed and highly complex. It is expected to be more demanding. It requires learning and rethinking gender power relations. In relation to security and defense governance, it requires learning about how gender is related to masculinity, violence, and militarism. This analysis has shown that gender parity seems far more difficult to achieve in the context of military and defense relations than gender mainstreaming. An explanation may be that in the context of military and defense relations, the presence of the male body is so closely tied to military practice that it can hardly be untangled from men, for example, when new institutions are built. The presence of the female body may be something far more provocative to military practice and organizations than the call to a depoliticized concept of gender or gender mainstreaming. It is also clear from the analysis that the concepts of gender and gender mainstreaming have been subject to a reinterpretation in the context of the ESDP. The process of reinterpreting what gender means is likely a part of its success.

The tendency in the EU's approach to UN SCR 1325 and 1820 is, in relation to training in ESDP missions, to fall back to the meaning of gender as woman and the relationship to local women particularly and to tie gender closely to operational effectiveness. While this is a way to allow military staff to grasp the meaning of and accept gender mainstreaming as a way to give added value to the operation, it reduces gender and UN SCR 1325 and 1820 to something instrumental. It is only when gender can be made operational in the specific task of the missions that it has relevance. Through this overview we have seen that gender has become a concept open to many interpretations (cf. Woodward 2008, 295). Gender as it has been applied in the ESDP is a concept completely deprived of power relations, and it is a concept of gender that has become "trapped in a heteronormative binary of man/woman" (Zalewski 2010, 22).

CONCLUDING COMMENTS

The aim of this chapter was to investigate whether the ESDP can be considered a postnational military and whether and how the ESDP has been influenced by the existing gender governance already in place in the EU and by the member states. In some respects the ESDP can be considered a postnational military. The battlegroups signify a shared commitment by the member states, and many are multinational, such as the Nordic Battlegroup. Furthermore, their mission is to 'save distant others' in the name of human rights, which would be characteristic of a postnational and even a cosmopolitan military. However, to be a full-fledged cosmopolitan military requires that the military organization also incorporate those ideals in its own organization, for example, by respecting gender equality norms and adhering to gender mainstreaming. The ESDP as an institutional arrangement is completely dominated by men, in leadership positions as well as in the troops. Clearly, gender parity has not been the ambition in setting up the new institutions; instead, national militaries of the member states have been able to entirely influence this and hence weaken the cosmopolitan element. The example of the Nordic Battlegroup indeed shows how militarism, masculinity, and sexuality are connected while also revealing a tension between traditional views on militarism and the emergence of a new EU defense arrangement. The same example also verifies the difficulty in achieving gender parity in military-type organizations. At the same time, and quite surprisingly, it shows that gender mainstreaming is more successful if taken to mean education and learning about gender effects and how gender can make the organization more operational. This goes against what has been commonly argued about gender mainstreaming as a much more difficult strategy to implement than overcoming the gender gap. It is related to the way the concept of gender has been interpreted and conceptualized in the ESDP as mainly signifying 'vulnerable women in faraway places'. Gender has been reinterpreted through the work in ESDP with UN SCR 1325 and 1820, to be viewed in an instrumental way as something providing added value and increasing operational effectiveness in the missions' operations. Gender has been incorporated into the ESDP, but at the same time the concept of gender has been emptied of politics and power.

CHAPTER 6

Conclusions

In responding to the main question presented in this book—what role does gender and sex play in the postnational defense?—we can conclude that gender relations associated with the dichotomy of protector and protected, an integral part of the national defense, has been called into question, substantially challenged, and to some degree reshaped. The following sections capture these transformations in relation to the most important themes of this book: They discuss the way the United Nations (UN) Security Council Resolution (SCR) 1325 has influenced peacekeeping practices, sketch the reconceptualization of women in the military as peacekeepers, outline the main conclusions on the role of sexuality and militarism in the postnational defense, and discuss the cosmopolitan imperative for postnational militaries.

UN SECURITY COUNCIL RESOLUTION 1325

In the postnational defense, there has been a shift in that the main emphasis is on human rather than national security. Considerations for human security were traced in the Swedish defense as well as other Nordic and European defense organizations in the EU and in Canada. Human security issues were articulated as one of the most important motivations for these defense organizations. The 'responsibility to protect' is one example of how international peacekeeping activities have become more human rights oriented; another is the UN Security Council's agenda on women, peace, and security, with Resolution UN SCR 1325 in 2000 as the first in a series of resolutions. The increasing number of resolutions on gender security issues can be interpreted as an increasing weight of gender aspects on the UN Security Council. An alternative rendition would be that the new resolutions are needed because of inadequate recognition and implementation of the original UN SCR 1325 among UN member states. Indeed, UN SCR 1820, 1888, and 1889 are meant to strengthen implementation of what is already covered in SCR 1325. Hence, while the question of whether the broader recognition of gender and security issues has influenced the international security agenda can be answered in the affirmative, the extent of its influence is more questionable.

In this study, the main concern has been to review how such gender security concerns have influenced the postnational defense and in particular the military organizations. It has been illustrated that ideas generated in the women's movement embedded in the resolutions have traveled to postnational military practices and on its way encountered and clashed with institutionalized norms of militarism. Thus, the study shows how gender as a 'traveling concept' has taken different expressions and meanings through the implementation of SCR 1325 in the military organization. To date, 25 countries have responded with National Action Plans; the Nordic countries, some European Union (EU) member states, and the EU are among them.[1] Sweden is on its second plan (which also includes 1820) and shows the most extensive integration of gender-security thinking, as exemplified with peacekeeping activities in the International Security Assistance Force (ISAF) in Afghanistan and in the Nordic Battlegroup. The process has included sensitizing military staff to the special security needs of women in conflict zones in the region of deployment, trying to find ways to involve and collaborate with women in the local context, and trying to anticipate effects on women in the military operational setting. In this respect, it has been no small feat to take a resolution formulated in the Security Council and let that trickle down to the concrete, daily practices of a provincial reconstruction team in Mazar-i-Sharif in Afghanistan. Indeed, as the book describes, it did require extensive efforts, including educational commitments, gender coaching, and extensive training throughout the entire organization. It required organizational innovations such as gender field advisors as well as a committed leadership that took responsibility for gender concerns. These were the success factors that the study points out.

The postnational cosmopolitan military, as is evident from the discussion of the peacekeepers in Afghanistan in chapter 4, is one that notices women; it sees and recognizes that women are victims in conflict areas and have different security needs. It is a military that realizes that women should be included when there are negotiations between conflicting parties or when different groups are cooperating to build peace, because their inclusion will increase the likelihood of the mission succeeding. The attention paid to gender issues is an indication of a sea change, particularly when these concerns come into the organization's operational level and into daily practices, because it means that a discussion is raised and people are forced to reflect on gender issues. There is much to learn from the Swedish example on how UN SCR 1325 can become operational in a postnational context and in international missions in particular. Many of the other military organizations, discussed here as cosmopolitan minded, are also looking to Sweden for examples on how this can be done.

Having said that, this study also shows that despite the fact that UN SCR 1325 calls for both increased gender balance and gender mainstreaming in peacekeeping forces and in the relations with the local community, what has become the main focus is the security situation of women in the conflict zone. While this is both an admirable and a crucial concern, the tendency is nevertheless to turn the broader gender perspective into a narrow concern for women as victims in conflict and zones of insecurity. While the EU's security and defense policy (ESDP) is the most obvious development in this direction, it is a general inclination of cosmopolitan-minded militaries when they try to conform to UN SCR 1325. *Gender* becomes equated with *woman* and materializes as a concern for and a focus specifically on the local women in the area where missions are based or are expected to be deployed.

In the case of the Swedish ISAF mission to Afghanistan, this led to a one-sided view of Afghan women. Male peacekeepers were advised not to talk to the women. Hence, the women of 'the other' were defined as all being equally powerless, subjected to the traditional patriarchal power of Afghan society. Not talking to women was a way of avoiding putting Afghan women at risk and recognizing the circumscribed agency that many women have in the Afghan society. Yet it reduced women to victims only. In such a context, if UN SCR 1325 is to be taken seriously, the military organization deployed in Afghanistan is dependent on female staff, the only channel by which to reach local women. It is difficult to include or even envision women who are perceived as victims as active agents in collaborations and peace negotiations. Female peacekeepers are considered the only ones who can discover the local women's security needs, include them in security and peace negotiations, and acquire intelligence information from them. Recalling that women continue to be a small minority in the peacekeeping forces, it is likely that only a small number of local women will be approached by the peacekeeping forces, and only a few have a chance to be included in security talks and peace negotiations. Furthermore, it may restrict female military staff's professional options in the mission because they are so badly needed to talk to the women.

When the concept of gender has traveled it has come to mean *woman*. Moreover, the main relevance of the gender/woman concept is related to the 'other' women to be found in the local community where missions are deployed. Evidently this interpretation of gender pays scarce attention to how relations of gender and sexuality are part of the peacekeeping practices within the mission. Questions about masculinity and sexuality are silenced. The postnational defense practices hold an interpretation of gender far from what has been intended in feminist theory, gender studies, and the woman's movement. It is the result of the concept's travels through institutions formed by militarism and heterosexual masculinity. Politics and power have been removed from the concept along the way, and it has turned gender mainstreaming into a problem-solving rather than a critical tool.

WOMEN IN THE MILITARY AS PEACEKEEPERS

The picture looks somewhat different for the postnational military in the domestic context. We have seen that postnational and cosmopolitan militaries are democratizing and trying to address gender and sexuality issues. For the Swedish military, we noted how this affected the organization internally through gender equality norms, the new values of the defense, and zero tolerance for discrimination and harassment. The call to gender mainstream military practices, as voiced by UN SCR 1325, is, however, only partially taken to heart. Surprisingly perhaps, the most marked challenge for the postnational defense is not to introduce gender perspectives in training—a form of gender mainstreaming and something that has been considered the more difficult issue in other policy fields. Rather, it has been to increase the share of women among the military staff and among its peacekeepers. To add women to the postnational military remains difficult, yet, as we have seen, it is absolutely necessary to reach some of the objectives set out in UN SCR 1325 and related resolutions. Women's presence in peacekeeping operations and in military institutions remains very low. The few women who nevertheless do join have difficulty in identifying themselves with and becoming a part of the organization.

Despite efforts on behalf of the military organization, the 'woman-in-arms' subjectivity also rests uneasily with the postnational defense. The postnational defense seems divided in how it views women. On the one hand, it works with the wishful thinking that women are soldiers and peacekeepers on the same terms and are just 'one of the boys', while at other times and in other contexts it emphasizes difference. This occurs as women are argued to be an added resource to the military organization with complementary skills and thus can be used for different tasks than men. For example, female peacekeepers are expected to install legitimacy to peacekeeping troops simply by their presence. In cases where male peacekeepers had entered into sexually exploitative relations with local women, female peacekeepers were the antidote—the ones who could put a damper on the sexuality of male peacekeepers and restore trust for the mission. Until the role of masculinity in relation to militarism and for military practice is deconstructed by the organization, the prospect that women will want to join and become soldiers or peacekeepers remains limited. This has validity beyond the Swedish Armed Forces, the EU Battlegroups, and the EU. Many European countries have relied on all-male conscription systems, and their nations have been gendered in a similar way. Women's presence in militaries throughout Europe is extremely low due to this conscription system, and this has largely been carried over to the new security and defense organization of the EU. Gender perspectives deeply challenge institutions of hegemonic masculinity that have evolved also into a postnational form.

MILITARY INSTITUTIONS AND SEXUALITY

As stated in chapter 2, military organizations as institutions of hegemonic masculinity imply that heterosexual masculinity has completely infused the organizations and become the norm. Heterosexual masculinity has been intimately related to how tasks and training are performed and to the conceptualization of an efficient military operation relying specifically on homosocial bonds among heterosexual men. Military training has built a specific masculinity that fits this imagery, often by using sexuality as an organizational resource. Despite this, this study argues that male sexuality was conceptualized in peacekeeping missions as strictly an individual matter in which male peacekeepers were not attributed any real agency or power over their sexuality. Sexual misconduct, for example, is perceived as the result of an innate drive. This is problematic because sexuality is an integral element in the military as an institution and key in its construction as a heterosexual, masculine organization. This is also expressed when that same organization is involved in peacekeeping activities in international missions.

The role of sexuality seems to be the least explored question of gender politics in the postnational defense. As argued in chapters 1, and 2, there has been a complete silence on gender and sexuality in the national military. The gendered dichotomies of the 'neutral soldiers' and 'mothers' made the military a place for heterosexual men, but neither their gender nor their sexuality has been a topic of discussion. In this sense, a lot has changed when a soldier is confronted regarding his homophobic jargon or forced to ask himself why he is the one who has to learn how to kill. The examples brought up in this book show that such debates are actually taking place within the cosmopolitan military organization. However, at times this may be simply rhetorical, as a way to respond to political concerns or appeal to the public.

Tolerance for sexual diversity around the 'new values' has been part of this process, but the strategies have been different from those we have seen for gender issues. The military has not articulated a wish to include more lesbian, gay, bisexual, and transgender (LGBT) persons in the military, nor have LGBT persons been conceptualized as having specific competences. The issue has been framed as a matter of respect for the rule of law. Sexuality remains constructed as individual and private and not in organizational terms.

Discrimination and harassment on the grounds of sexual preferences is not permitted in the EU. Nevertheless, in the postnational defense, the tolerance for sexual diversity becomes a rather complicated issue in practice. This is in part due to the importance attributed to homosociality and military performance but also because tolerance is not as widespread, or as formalized, as the acceptance for gender approaches. There are no specific UN resolutions protecting LGBT persons. Furthermore, the view on homosexuality varies among multilateral forces as well as in different local contexts. To be open about one's homo-, bi-, or transsexuality as a peacekeeper may entail a severe security risk, particularly in locations where such preferences are considered a crime or even a capital offense. In the postnational defense studied, there is a silenced tolerance. In this way, European militaries differ very much from the U.S. military, where there has been a long and sometimes heated debate on whether to tolerate homosexuals in the military. In Europe, the issue can be characterized by quiet (in)tolerance. This difference can perhaps be explained by the long tradition of using conscription to staff the military; conscripts are not part of the regular labor market and have not been subject to or able to draw on occupational rights in the same way as would be the case for a professional military. Difference in sexual preference is tolerated, and there is some basic understanding, at least in Sweden and among some EU member states, that a state military cannot do otherwise but tolerate. However, the military is not discussing what hetero-, bi-, homo-, or transsexuality implies for military practice, peacekeeping, or the organization. Can it be a resource for the postnational defense? Or is it a matter of representing the full range of citizens?

MILITARISM AND THE GENDER ORDER

Feminist international relations research has highlighted the importance of militarism as a norm system that motivates the military's existence and considers militarism as closely connected to the gender order. The cases studied here contribute to the understanding of militarism and its relation to the gender order. They show that militarism is not necessarily connected to a high degree of violence but can also be latent yet deeply embedded in society. Sweden had a neutral defense with a long history without war. The kind of militarism that developed involved almost every citizen in one way or other in the nation's defense. The neutral defense was territorially based but also closely linked with nation building. Thus, a gendered militarism permeated conscription, military institutions, political bodies, and the voluntary defense organizations. This specific militarism ran like a thread throughout the nation into every home via conscripts. It evolved around a range of duties, the handling and production of military equipment of various kinds, and the development of defense strategies but also via the voluntary defense organizations with tasks such as driving motorcycles and heavy vehicles, cooking pea soup, and milking cows. These tasks linked the Swedish citizens to the defense of the nation, a link or thread that unraveled in the postnational defense.

Another example of how militarism can be a latent phenomenon was illustrated through the EU case. A civilian type of organization incorporates a concept such as a battlegroup—a military term—and the institutionalization of EU defense activities relies almost exclusively on military expertise, staff, and ideas from the member states' national defense. It is exemplified in the EU Military Committee. These analyses of the ESDP show how militarism as a norm system, without much reflection, becomes part of a new institutional setup. That it is done without reflection is an indication that militarism is embedded and normative for defense institutions and thinking in Europe; hence, it is simply transferred to a new institutional arrangement such as the battlegroups, the Military Committee, and the ESDP.

THE POSTNATIONAL DEFENSE AND COSMOPOLITAN VALUES

This book took as its starting point the broader recognition that human security's salience in global relations has influenced the way the defense and military are organized. A useful way to probe this issue was through Elliott and Cheeseman's (2004) theory on cosmopolitan militaries. It was particularly interesting to study the potential for reform, embedded in cosmopolitanism as an ethic for militaries, from a feminist perspective. First, the focus was on the postnational defense that questions national and territorial identification notable in the imperative to show 'solidarity with distant others'. The notion that solidarity should be with distant others is a radical plea that calls into question the basic imperative of the military. It tests the relation between the military and the national project by expecting sacrifices that are made in the name of human security for people far away with no immediate connection to the nation except that they belong to humanity and deserve dignity and respect.

Although the focus has been on Sweden as an example of a cosmopolitan military, Sweden's experiences have been related to how other militaries of the postnational defense have developed in the rest of Europe and within the EU. Such militaries, I argue, distinguish themselves from militaries like those of the United States and Israel, nonstate militaries engaged in civil wars or ethnic conflicts, and privatized militaries. What sets them apart is that they have a human security focus, and this is the position they take in the international security context. The book analyzed what demands the cosmopolitan imperatives put on these militaries.

To think in terms of cosmopolitan militaries was intriguing also because cosmopolitan values hold the potential to demilitarize the military. This is due to its emphasis on peace making rather than war. The demand to work for human security in peace enforcing or peacekeeping operations has forced militaries to become more diversified in the range of skills they use. The tasks that a cosmopolitan military is expected to master range from combat to communication. It is a formidable challenge to turn an organization with a central and professional capability tied to skills around the use of violence in combat and combat-like activities, using weapons that kill, to an organization that does peace.

For the postnational defense, a pertinent query is whether the changing tasks of 'doing peace instead of war' also is a way to demilitarize the military defense organization. The cosmopolitan military, it is argued, tends to prefer using means other than military ones, which often are also more suitable to peacekeeping operations. International and EU missions as well as the EU Battlegroups require that the military extend its set of abilities to a whole range of means to succeed in its mission. This entails everything from developing a dialogue with local inhabitants, to helping meet the daily needs of the population, to training the local police forces, to securing areas and buildings, sometimes with military force. Yet what distinguishes militaries is their training and skill in the use of military violence and equipment. The study showed how military practices are closely tied to this skill and to the performance of masculinity. The changing tasks of cosmopolitan militaries also raise questions about masculinity's relation to violence and the performance of peacekeeping duties.

Another argument is that a cosmopolitan military, to be true to its commitment, must also reform its organization to adhere to democratic norms in society (i.e., respect equality and sexual differences; forbid discrimination and any kind of harassment). The Swedish postnational military has such an ambition. Reforms to include democratic norms and respect for law and human rights has been a part of the military's internal strategies for some time. The main conclusion from this study is that despite many good intentions and aspirations along these lines, there are many obstacles for militaries attempting to take a cosmopolitan-minded path. It is explained by the way that militarism and masculinity is at the core of military organizations and defense thinking.

A FEMINIST POSTNATIONAL DEFENSE?

As this book ends, there remains an unresolved ambivalence around the use of force and violence worthy of discussion. On the one hand, a cosmopolitan military must appear militarily prepared to be trustworthy and useful. At the same time, if it has to use this force, it can jeopardize trust and legitimacy. The most challenging issue for postnational cosmopolitan militaries is when and on what grounds violence can be used. This presents an ethical dilemma for a cosmopolitan military. A similar ethical dilemma confronts feminists.

To give women agency in postnational defense matters is not simply about representation and increasing the number of female bodies in a setting dominated by male bodies. It is about critically looking at gender relations. This implies deconstructing, unraveling, and critically exploring the relations among militarism, masculinity, violence, and sexuality. One implication of doing so is that feminists can no longer position themselves above or outside questions of militarism and defense. Is pacifism the only alternative to antimilitarism? In my view—and here I follow the thoughts of Laura Sjoberg (2006) in her feminist reformulation of Just War Theory—women and feminists must consider the question of whether there are times when violence might actually be necessary. This position does not entail a glorification of violence or an acceptance of militarism but it includes a recognition that the use of violence may in some cases be the only possible option. To use organized violence would be the last resort, not a desirable solution. It is possible to imagine that atrocities committed are so intolerable that women, too, would be willing to die in an attempt to put an end to suffering. We know already from other scholarly work that women are capable of committing violence and take part in military activities. It is possible to vision a women-in-arms subjectivity with the agency to sacrifice herself for values that she endorses. This way women have influence in military affairs and defense matters, but it also means that women cannot be innocent. They can no longer represent peaceful and beautiful souls but must take responsibility for global relations in the field of war economy, militarism, and defense. The lack or very limited agency that women have in the defense field, which this analysis has pointed out, implies that women have in practice also been barred from the opportunity to provide alternative security and defense policies in international relations and diplomacy.

If feminists and women activists see organized violence as necessary in specific cases and circumstances as described above, then it begs the question: What type of postnational defense does this call for? What would alternative security and defense politics look like? Pushing the question further: Would a feminist state also need a defense that has as a core skill the ability to use violence and combat to achieve human security? Cosmopolitan values as they have been discussed here can help define such an organization but do not suffice. In chapter 3 I proposed that a feminist international security ethic based on the ethics of care and emphatic cooperation could be a way to guide the postnational defense in difficult considerations and choices in the future when dealing with establishing security, keeping peace, managing crises, and upholding law and order in places where there have been violations of human rights and sensitivity to the local context is necessary. While the organization must be prepared to use violence, military practices and combat skills must surely be decoupled from masculinity. Civilian tasks and functions become more important and well integrated into the organizations' overall activities. Ethics of care and emphatic cooperation are highly valued skills in engaging with the local populations. Some important features of military organizations, such as their trained capacity to take immediate action on specific tasks, could be put to good use in quickly securing areas, upholding law and order, and assisting local people in different tasks related to building the conditions for peace. As a key feature of the feminist postnational defense organization, militarism would be reconceptualized away from combat to the ability to take action to protect 'distant others' so that they may live to their full human potential.

NOTES

INTRODUCTION
1. For more general information on ISAF see http://www.isaf.nato.int.
2. UN SCR 1386 (2001), UN SCR 1510 (2004), and UN SCR 1833 (2008).
3. Two projects conducted together with Erika Svedberg of Örebro University have been particularly important in this regard. The first project was on conscription practice in Sweden, financed by the Swedish Research Council during 1999–2003. The second project studied the voluntary defense organizations and was financed by the Swedish Crises Management Agency, with the results presented in an internal research report in Swedish.

CHAPTER 1
1. Parental leave is part of the Swedish welfare system, and one parent can stay home with a child for the first 16 months with employment guaranteed and the wage paid (up to a certain pay ceiling) during the entire period. Although an option for mothers or fathers, mothers use the largest share of parental leave within the family.
2. What is argued here on the relationship between conscription and citizenship for Sweden is equally relevant for Denmark, Finland, and Norway.
3. They have, however, not been completely inactive in security politics. Women politicians have pursued foreign policy issues connected to the UN and development aid.
4. Many of these voluntary defense organizations are still active, but they do not work under the same conditions as during neutrality. Membership has diminished, state funding has been cut, and conditions and tasks have changed drastically since the end of the Cold War. Their role toward nation-making has diminished in importance but is not analyzed in this book.
5. In Swedish, the term *voluntary organization* is also used for *civil society organizations* and nongovernmental organizations in general.

CHAPTER 3
1. There is an extensive literature on cosmopolitanism with many debates and criticisms that I do not take into account in this book. I restrict this study mainly to the use and application of the concept of cosmopolitan militaries because it points to and exemplifies some of the elements that are central to a feminist analysis of gender, sex, and the postnational defense.
2. Sweden differs from Switzerland, Austria, and Finland, whose neutrality was bound by international convention and bilateral agreement.

3. For example, Olof Palme UN Commission on Disarmament and Security, 1982.
4. In the perspective for the future report prepared by SAF headquarters, it is articulated as: "The equality issue will be an important human resources issue in the future organization. Other nations have not taken equality seriously" (my translation; see PERP (2009, 32).
5. Military scholar Van Creveld (2008, 2001) openly expresses the view that a feminization is highly detrimental to the military. In his view, both demilitarization and demasculinization would contribute to feminization and reduce military effectiveness.
6. There is much research on posttraumatic stress disorder, and the dimension discussed by Whitworth (2004) is only one aspect of this.
7. This is particularly stressed on the SAF's website informing on equality issues. See *Försvarsmaktens jämställdhets -och jämlikhetsarbete.* 2009.
8. http://www.mil.se/sv/Forband-och-formagor/Forband/Livgardet-LG/Nyheter/ Sverige-leder-unik-genderkurs/. Accessed October 8, 2009.
9. http://www.forsvarsmakten.se/sv/Om-Forsvarsmakten/Forsvarsmakten-i-fickformat-2009/Accessed August 30, 2011.
10. http://www.forsvarsmakten.se/templates/Mil_UnitStartpage.aspx?id= 20475&epslanguage=EN. Accessed April 16, 2011.

CHAPTER 4

1. I use *peacekeeping* as a generic term for all international missions of the UN, EU, and NATO.
2. UNIFL: http://www.un.org/en/peacekeeping/missions/unifil/. Accessed April 27, 2011.
3. I conclude this based on my general research in the field but also confirmed in a response to a direct question regarding whether masculinities have been discussed in the military to Major General Bengt Andersson, Chief Logistic Officer of the SAF, and Colonel Jan Blacquiere, Head of Current Operations, Dutch Defense Staff at the Second Swedish-Dutch Conference on Gender Equality: Women in War Zones, December 3, 2009, Peace Palace, The Hague.
4. http://www.avert.org/age-of-consent.htm. Accessed December 15, 2009.
5. http://www.un.org/en/peacekeeping/contributors/gender.shtml. Accessed June 15, 2010.
6. The argument is that diversity and gender are resources. See *Försvarsmaktens värdegrunder* (2006).
7. See also the 1995 Beijing platform of action, Objective E4, which concerns this http://www.un.org/womenwatch/daw/beijing/platform/armed.htm. Accessed August 30, 2011.
8. http://www.forsvarsmakten.se/sv/Om-Forsvarsmakten/Verksamhet/Internationellt_arbete/Genderforce/. Accessed August 30, 2011.
9. http://www.forsvarsmakten.se/sv/Internationella-insatser/. Accessed August 30, 2011.

CHAPTER 5

1. The analysis of ESDP is limited to gender conceptualizations, gender parity, and mainstreaming because of limits in data. There is no research or data available for the ESDP related to diversity, sexuality, or harassment issues.
2. http://www.consilium.europa.eu/policies/foreign-policy/eu-special-representatives. aspx?lang=en. Accessed April 3, 2011.

3. http://www.consilium.europa.eu/eeas/security-defence/csdp-structures-and-instruments/eu-military-committee-(eumc).aspx?lang=en. Accessed April 3, 2011.
4. Member state reports to questionnaire (Swedish Presidency 2009).
5. The SAF have run an advertisement campaign in the media that conveys a combat-intensive view of its military practices in a manner not unlike a Hollywood film. The ad received criticism for not giving an accurate or serious picture of what the Swedish military actually does on international missions and because it may actually attract the kind of trigger-happy recruits it does not want. See, for example, *Dagens Nyheter*, June 15, 2010. http://www.dn.se/nyheter/sverige/forsvarskampanj-ifragasatt-1.1122289. Accessed June 25, 2010.
6. http://www.consilium.europa.eu/homepage.aspx?lang=en. Accessed March 31, 2011.
7. http://www.db-decision.de. Accessed June 17, 2010.
8. http://www.genderforce.se/dokument/From_words_to_action.pdf. Accessed April 2, 2011.

CHAPTER 6
1. http://www.peacewomen.org/pages/about-1325/national-action-plans-naps. Accessed March 15, 2011.

REFERENCES

Acker, Joan. 1992. "Gendering Organizational Theory." In *Gendering Organizational Analysis*, ed. Albert J. Mills and Peta Tancred, 248–260. Newbury Park, CA: SAGE.

Ackerly, Brooke, Maria Stern, and Jacqui True, eds. 2006. *Feminist Methodologies for International Relations.* Cambridge, UK: Cambridge University Press.

Agathangelou, Anna, and L.H.M. Ling. 2003. "Desire Industries: Sex Trafficking, UN Peacekeeping, and the Neo-Liberal World Order." *Brown Journal of World Affairs*, 10(1): 133–148.

Aggestam, Karin, ed. 2004. *(O)rättfärdiga krig* [(Un)just wars]. Lund, Sweden: Studentlitteratur.

Agrell, Wilhelm. 1991. *Den stora lögnen-Ett säkerhetspolitiskt dubbelspel i alltför många akter* [The great lie—A security game]. Stockholm: Ordfronts förlag.

———. 2000. *Fred och fruktan. Sveriges säkerhetspolitiska historia 1918–2000* [Peace and fear. Sweden's security policy history, 1918–2000]. Lund, Sweden: Historiska media.

Åkerström, Linda. 2008. *Säkerhetspolitik för en ny tid* [Security policy for a new time]. Stockholm: Pax.

Albanese, Patrizia. 2006. *Mothers of the nation: Women, families, and nationalism in twentieth-century Europe.* Toronto: University of Toronto Press.

"Allt farligare i Afghanistan." 2005. *Dagens Nyheter*, November 25.

Ambjörnsson, Ronny. 1998. *Den skötsamme arbetaren* [The conscientious worker]. Stockholm: Carlsson.

Anderson, Benedict. 1983. *Imagined communities: Reflections on the origin and spread of nationalism.* London: Verso.

Anderson, Letitia. 2010. "Politics by Other Means: When Does Sexual Violence Threaten International Peace and Security?" *International Peacekeeping*, 17(2): 244–260.

Andersson, Jan Joel. 2006. "Armed and Ready? The EU Battlegroup Concept and the Nordic Battlegroup." *SIEPS* 2006:2. Stockholm: Swedish Institute for European Policy Studies.

"Ännu en svensk Afghanistansoldat död." 2005. *Dagens Nyheter*, December 9.

Armadillo. 2009. Directed by Janus Metz Pedersen. Mosedalvej 14, Denmark: Nordisk Film Studio.

"Army Castrates Heraldic Lion." 2007. *The Local: Sweden's News in English*. December 13.

Åselius, Gunnar. 2005. "Swedish Strategic Culture after 1945." *Cooperation and Conflict*, 40(1): 25–44.

Åström, Sverker. 1989. "Swedish Neutrality: Credibility through Commitment and Consistency." In *The Committed Neutral: Sweden's Foreign Policy*, ed. Bengt Sundelius, 15–33. Boulder, CO: Westview.

Atterling, Susanne, Vera Cadova, Robert Nises, and Börje Rosmark. 2001. *Värnpliktiga kvinnor i fokus*. [Conscripted women in focus]. Stockholm: Pliktverket.

Bacevich, Andrew J. 2006. *The New American Militarism: How Americans Are Seduced by War*. New York: Oxford University Press.

Bailes, Alyson J.K. 2006. "Introduction: The European Defence Challenge for the Nordic Region." In *The Nordic Countries and the European Security and Defence Policy*, ed. Alyson J.K. Bailes, Gunilla Herolf, and Bengt Sundelius, 1–26. Stockholm: SIPRI.

Baker, Henderson. 2006. "Women in Combat—A Cultural Issue?" Master's thesis, U.S. Army War College, Carlisle, PA.

Barkawi, Tarak, Christopher Dandeker, Melissa Wells-Petry, and Elizabeth Kier. 1999. "Rights and Fights: Sexual Orientation and Military Effectiveness." *International Security*, 24(1): 181–202.

Barrett, Frank J. 2001. "The Organizational Construction of Hegemonic Masculinity: The Case of the U.S. Navy." In *The Masculinities Reader*, ed. Stephen Whitehead and Frank J. Barrett, 77–99. Cambridge, UK: Polity.

Barth, Elise. 2004. "The United Nations Mission in Eritrea/Ethiopia: Gender(ed) Effects." In *Gender Aspects of Conflict Interventions: Intended and Unintended Consequences*, ed. Louis Olsson, 9–24. Oslo: International Peace Research Institute.

Bateman, Geoffrey, and Sameera Dalvi. 2004. *Multinational Military Units and Homosexual Personnel*. Santa Barbara: University of California, Center for the Study of Sexual Minorities in the Military.

Beck, Ulrich. 2006. *The Cosmopolitan Vision*. Cambridge, UK: Polity.

Ben-Eliezer, Uri. 1998. *The Making of Israeli Militarism*. Bloomington: Indiana University Press.

Berdahl, Jennifer. 2007. "Harassment Based on Sex: Protecting Social Status in the Context of Gender Hierarchy." *Academy of Management Review*, 32(2): 641–658.

Berggren, Anders W., and Sophia Ivarsson, eds. 2002. *Jakten sätter på attacken* [On sexual harassment in SAF]. Stockholm: Försvarshögskolan.

Bergman, Annika. 2007. "Co-Constitution of Domestic and International Welfare Obligations: The Case of Sweden's Social Democratically Inspired Internationalism." *Cooperation and Conflict*, 42(1): 73–99.

———. 2006. "The Concept of Solidarity and Post–Cold War Nordic-Baltic Relations." *Cooperation and Conflict*, 41(1): 73–97.

———. 2004. "The Nordic Militaries: Forces for Good?' In *Forces for Good? Cosmopolitan Militaries in the Twenty-First Century*, ed. Lorraine Elliott and Graeme Cheeseman, 168–186. Manchester, UK: Manchester University Press.

Bildt, Carina. 2004. "Arbetslivsinstitutets redovisning av regeringsuppdraget att beforska homo- och bisexuellas arbetsvillkor" [The working conditions for LGBT]. *Arbetslivsrapport* NR 2004:16. Stockholm: Arbetslivsinstitutet.

Bird, Sharon R. 1996. "Welcome to the Men's Club. Homosociality and the Maintenance of Hegemonic Masculinity." *Gender & Society*, 10(2): 120–132.

Biricik, Alp. 2011. "The 'Rotten Report' and the Reproduction of Masculinity: Nation and Security in Turkey." In *Making Gender, Making War: Violence, Military, and Peacekeeping Practices*, ed. Annica Kronsell and Erika Svedberg, 76–89. New York: Routledge.

Bjelanovic, Vera. 2004. *Fröken Duktig som Fänrik Karsk. En uppföljning av Pliktverkets studie Värnpliktiga kvinnor i fokus* [Focus on female conscripts]. Pliktverkets rapportserie, rapport 13. Östersund, Sweden: Pliktverket.

Björkdahl, Annika. 2002. "From Idea to Norm—Promoting Conflict Prevention." Ph.D. diss., Lund University, Lund, Sweden.

———. 2008. "Norm Advocacy: A Small State Strategy to Influence EU." *Journal of European Public Policy*, 15(1): 135–154.

———. 2005. "Peace Operations and the Promotion of Cosmopolitanism." *Statsvetenskaplig Tidskrift*, 107(3): 215–233.

Blomgren, Ebbe, and Ove Lind. 1997. *Kvinna som man är*, LI Serie T:2. Stockholm: Försvarshögskolan, Ledarskapsinstitutionen.

Bolin, Anna. 2008. "The Military Profession in Change: The Case of Sweden." Ph.D. diss., Lund University, Lund, Sweden.

Bridges, Donna, and Debbie Horsfall. 2009. "Increasing Operational Effectiveness in UN Peacekeeping: Toward a Gender-Balanced Force." *Armed Forces & Society*, 36(1): 120–130.

"Brottslingar attackerar svensk styrka." 2006. *Dagens Nyheter*, September 8.

Buchanan, Allen, and Robert O. Keohane. 2004. "The Preventive Use of Force: A Cosmopolitan Institutional Proposal." *Ethics & International Affairs*, 18(2): 1–22.

Burrell, Gibson, and Jeff Hearn. 1989. "The Sexuality of Organization." In *The Sexuality of Organization*, ed. Jeff Hearn, Deborah L. Sheppard, Peter Tancred, and Gibson Burrell, 1–28. Newbury Park, CA: SAGE.

Burrelli, David. 2010. *"Don't Ask, Don't Tell": The Law and Military Policy on Same-Sex Behavior*. Washington, DC: Congressional Research Service.

Butler, Judith. 2009. *Frames of War: When Is Life Grievable?* London: Verso.

———. 1990. *Gender Trouble: Feminism and the Subversion of Identity*. New York: Routledge.

———. 2004. *Precarious Life: The Powers of Mourning and Violence*. New York: Verso.

Carlton, Eric. 2001. *Militarism—Rule without Law*. Aldershot, UK: Ashgate.

Carreiras, Helena. 2008. "From Loyalty to Dissent: How Military Women Respond to Integration Dilemmas." In *Women in the Military and in Armed Conflict*, ed. Helena Carreiras and Gerhard Kümmel, 161–181. Wiesbaden, Germany: VS Verlag für Sozialwissenschaften.

Carreiras, Helena, and Gerhard Kümmel. 2008. "Off Limits: The Cults of the Body and Social Homogeneity as Discursive Weapons Targeting Gender Integration in the Military." In *Women in the Military and in Armed Conflict*, ed. Helena Carreiras and Gerhard Kümmel, 29–47. Wiesbaden, Germany: VS Verlag für Sozialwissenschaften.

Charlesworth, Hilary. 2008. "Are Women Peaceful? Reflections on the Role of Women in Peace-Building." *Feminist Legal Studies*, 16: 347–361.

Cockburn, Cynthia. 2011. "Gender Relations as Causal in Militarization and War: A Feminist Standpoint." In *Making Gender, Making War: Violence, Military, and Peacekeeping Practices*, ed. Annica Kronsell and Erika Svedberg, 19–34. New York: Routledge.

———. 2004. "The Continuum of Violence. A Gendered Perspective on War and Peace." In *Sites of Violence: Gender and Conflict Zones*, ed. Wenona Mary Giles and Jennifer Hyndman, 24–44. Berkeley: University of California Press.

Cohn, Carol, H. Kinsella, and Susan Gibbings. 2004. "Women, Peace and Security: Resolution 1325." *International Feminist Journal of Politics*, 6(1): 130–140.

Coker, Christopher. 2008. *Ethics and War in the 21st Century*. London: Routledge.

———. 2001. *Humane Warfare*. London: Routledge.

———. 2007. *The Warrior Ethos. Military Culture and the War on Terror*. London: Routledge.

Collinson, David L., and Margaret Collinson. 1989. "Sexuality in the Workplace: The Domination of Men's Sexuality." In *The Sexuality of Organization*, ed. Jeff Hearn, Deborah L. Sheppard, Peter Tancred, and Gibson Burrell, 91–109. Newbury Park, CA: SAGE.

Connell, R.W. 1995. *Masculinities*. Cambridge, UK: Polity.

———. 1998. "Masculinities and Globalization." *Men and Masculinities*, 1(1): 3–23.

Connell, R.W., and James Messerschmidt. 2005. "Hegemonic Masculinity: Rethinking the Concept." *Gender & Society*, 19(6): 829–859.

Cooke, Miriam. 1996. *Women and the War Story*. Berkeley: University of California Press.

Council of the European Union. 2008a. *Comprehensive Approach to the EU Implementation of the UN SCR 1325 and 1820 on Women, Peace and Security* 15671/1/08. Brussels: Author. http://www.consilium.europa.eu/showPage.aspx?id=1886&lang=en. Accessed March 31, 2011.

———. 2008b. *Handbook: Mainstreaming Human Rights and Gender into European Security and Defence Policy*. Brussels: General Secretariat of the Council.

———. *Report on the Implementation of the European Security Strategy—Providing Security in a Changing World*. 2008c. Report S407/08. Brussels: Author. http://consilium.europa.eu/eeas/security-defence/european-security-strategy.aspx?lang=en. Accessed April 3, 2011.

———. 2008d. *Review of the Implementation by the MS and the EU Institutions of the Beijing Platform for Action, Conclusion on Indicators concerning Women and Armed Conflicts*, 17099/08. Brussels: Author.

Cronqvist, Marie. 2008. "Utrymning i folkhemmet. Kalla kriget, välfärdsidyllen och den svenska civilförsvarskulturen 1961." *Historisk Tidskrift*, 128(3): 451–476.

Dahlerup, Drude, and Lenita Freidenvall. 2008. *Kvotering*. Stockholm: SNS Förlag.

D'Amico, Francine, and Laurie Lee Weinstein, eds. 1999. *Gender Camouflage: Women and the U.S. Military*. New York: New York University Press.

Davis, Karen. 1997. "Understanding Women's Exit from the Canadian Forces: Implications for Integration." In *Wives & Warriors: Women and the Military in the U.S. and Canada*, ed. Laurie Weinstein and Christie White, 179–198. Westport, CT: Bergin & Garvey.

Defense Ministry. 1995. *Dokument från Konferens för kvinnliga officerare och aspiranter* [Documentation from conference with female officers]. Eskilstuna, April 5–6. Stockholm: Försvarsdepartementet.

DeGroot, Gerard. 2001. "A Few Good Women: Gender Stereotypes, the Military and Peacekeeping." In *Women and International Peacekeeping*, ed. Louise Olsson and Torunn L. Tryggestad, 23–38. Portland, OR: Cass.

"Den afghanska dödsdansen." 2009. *Dagens Nyheter*, January 29.

Denk, Thomas. 1999. *Värnpliktsutbildningen—en politisk socialisationsagent?* [Compulsory military training—A political socialization agent?]. *Karlstad University Studies*, 10.

"De vinkar inte, de drunknar." 2009. *Dagens Nyheter*, January 27.

Dexter, Helen, and Jonathan Gilmore. 2006. "Be Careful What You Wish For: Cosmopolitanism and the Renaissance of Warfighting." Centre for International Politics Working Paper Series 23. University of Manchester, Manchester, UK.

Dijkstra, Hylke. 2008. "The Council Secretariat's Role in the Common Foreign and Security Policy." *European Foreign Affairs Review*, 13(2): 149–166.

Diken, Bülent, and Carsten Bagge Laustsen. 2005. "Becoming Abject: Rape as a Weapon of War." *Body & Society*, 11(1): 111–128.

DiTomaso, Nancy. 1989. "Sexuality in the Workplace: Discrimination and Harassment." In *The Sexuality of Organization*, ed. Jeff Hearn, Deborah L. Sheppard, Peter Tancred, and Gibson Burrell, 71–90. Newbury Park, CA: SAGE.

Dixon, Erica. "Interview on Prostitution in Afghanistan." 2010. *Source*. August 16. http://www.slideshare.net/heatherrea/interview-on-prostitution-in-afghanistan. Accessed April 8, 2011.

Dohlman, Ebba. 1989. "Sweden: Interdependence and Economic Security." In *The Committed Neutral: Sweden's Foreign Policy*, ed. Bengt Sundelius, 95–121. Boulder, CO: Westview.

Dörfer, Ingemar. 1997. *The Nordic Nations in the New Western Security Regime*. Baltimore, MD: Johns Hopkins University Press.

Dowling, Colette. 2000. *The Frailty Myth: Women Approaching Physical Equality*. New York: Random House.

Ds 2003:8. 2003. *Säkrare grannskap—osäker värld* [Secure neighborhood—insecure world]. Stockholm: Försvarsberedningen, Försvarsdepartementet.

Ds 2004:30. 2004. *Försvar för en ny tid* [Defense in a new time]. Stockholm: Försvarsberedningen, Försvarsdepartementet.

Ds 2007:46. 2007. *Säkerhet i samverkan* [Security in cooperation]. Stockholm: Försvarsberedningen, Försvarsdepartementet.

Dudink, Stefan. 2002. "The Unheroic Men of a Moral Nation: Masculinity and Nation in Modern Dutch History." In *The Postwar Moment: Militaries, Masculinities and International Peacekeeping*, ed. Cynthia Cockburn and Dubravka Zarkov, 146–161. London: Lawrence & Wishart.

"Dutch Fury at U.S. General's Gay Theory over Srebrenica." 2010. BBC News, March 19. http://news.bbc.co.uk/go/pr/fr/-/2/hi/europe/8575717.stm.

Eduards, Maud. 2007. *Kroppspolitik. Om Moder Svea och andra kvinnor* [Body politics. Mother Svea and other women]. Stockholm: Atlas.

———. 2011. "What Does a Bath Towel Have to do with Security Policy? Gender Trouble in the Swedish Armed Force." In *Making Gender, Making War: Violence, Military, and Peacekeeping Practices*, ed. Annica Kronsell and Erika Svedberg, 51–62. New York: Routledge.

Ek, Anna. 2011. "Svenska Freds: de nya bombliberalerna en PR-succé för JAS." *Svenska Dagbladet*, March 22. http://svtdebatt.se/2011/03/svenska-freds-de-nya-bombliberalerna-en-pr-succe-for-jas/. Accessed March 22, 2011.

Elgström, Ole. 1982. "Aktiv utrikespolitik" [Active foreign policy]. Ph.D. diss., Lund University, Lund, Sweden. +

Elliott, Lorraine. 2004. "Cosmopolitan Ethics and Militaries as 'Forces for Good.'" In *Forces for Good? Cosmopolitan Militaries in the Twenty-First Century*, ed. Lorraine Elliott and Graeme Cheeseman, 17–32. Manchester, UK: Manchester University Press.

———. 2002. "Cosmopolitan Theory, Militaries and the Deployment of Force." Working Paper 2002/8. Department of International Relations, Australian National University, Canberra.

Elliott, Lorraine, and Graeme Cheeseman, eds. 2004. *Forces for Good? Cosmopolitan Militaries in the Twenty-First Century*, Manchester, UK: Manchester University Press.

Ellner, Andrea. 2008. "Regional Security in a Global Context: A Critical Appraisal of European Approaches to Security." *European Security*, 17(1): 9–31.

Elshtain, Jean Bethke. 1995. *Women and War*. Chicago: Chicago University Press. (Originally published 1987)

———, ed. 1992. *Just War Theory*. New York: New York University Press.

Engelbrektson, Karl. 2008. Interview at SAF Headquarters, Stockholm. September 24.

Engström, Mats. 2011. *Anna Lindh och det nya Europa.* Stockholm: Premiss.

Enloe, Cynthia. 1989. *Bananas, Beaches and Bases: Making Feminist Sense of International Politics.* Berkeley: University of California Press.

———. 2007. *Globalization and Militarism: Feminists Make the Link.* Lanham, MD: Rowman & Littlefield.

———. 2000. *Maneuvers—The International Politics of Militarizing Women's Lives.* Berkeley: University of California Press.

———. 2010. *Nimo's War, Emma's War: Making Feminist Sense of the Iraq War.* Berkeley: University of California Press.

———. 2004. *The Curious Feminist. Searching for Women in a New Age of Empire.* Berkeley: University of California Press.

———. 1993. *The Morning After: Sexual Politics at the End of the Cold War.* Berkeley: University of California Press.

Epstein, Debbie. 1997. "Keeping Them in their Place: Hetero/Sexist Harassment, Gender and the Enforcement of Heterosexuality." In *Sexual Harassment—Contemporary Feminist Perspectives*, ed. Alison M. Thomas and Celia Kitzinger, 154–171. Buckingham, UK: Open University Press.

Ericson, Lars. 1999. *Medborgare i vapen—värnplikten i Sverige under två sekel* [Citizens in arms—Conscription in Sweden over two centuries]. Lund, Sweden: Historisk media.

EU Security and Defence. 2005. *Core Documents 2004*, Vol. V. Chaillot Paper 75. Paris: Institute for Security Studies.

———. 2007. *Core Documents 2006*, Vol. VII. Chaillot Paper 98. Paris: Institute for Security Studies.

Eveline, Joan, and Carol Bacchi. 2005. "What Are We Mainstreaming When We Mainstream Gender?" *International Feminist Journal of Politics*, 7(4): 496–512.

Fahlstedt, Krister. 2000. *Studie över situationen för homosexuella inom Försvarsmakten, enskild uppsats inom Fackprogrammet i personalledning* [A study of the conditions for homosexuals in SAF]. Stockholm: Försvarsmakten.

Farwell, Nancy. 2004. "War Rape: New Conceptualizations and Responses." *Affilia*, 19(4): 389–403.

Ferguson, Kathy E. 1984. *The Feminist Case Against Bureaucracy.* Philadelphia: Temple University Press.

Fine, Robert. 2006. "Cosmopolitanism and Violence: Difficulties of Judgment." *The British Journal of Sociology*, 57(1): 49–67.

Fleming, Peter. 2007. "Sexuality, Power and Resistance in the Workplace." *Organization Studies*, 28(2): 239–256.

Flood, Michael. 2008. "Men, Sex, and Homosociality." *Men and Masculinities*, 10(3): 339–359.

Fogarty, Brian. 2000. *War, Peace and the Social Order.* Boulder, CO: Westview.

"Förhöjt hot mot styrkan i Afghanistan." 2005. *Dagens Nyheter*, December 19.

Försvarsberedningen. 2004. *Genusperspektiv på fredsbefrämjande insatser* [A gender perspective on peace-keeping missions]. Stockholm: Defense Committee. http://www.forsvarsberedningen.gov.se. Accessed June 13, 2005.

Försvarsmaktens jämställdhets -och jämlikhetsarbete. 2009. Stockholm: Försvarsmakten. www.mil.se/sv/Om-Forsvarsmakten/Arbetsplatsen/Jamstalldhetsarbete. Accessed November 28, 2009.

Försvarsmaktens värdegrunder [Values for the defense]. 2006. Stockholm: Försvars-
    makten http://www.forsvarsmakten.se/sv/Om-Forsvarsmakten/uppdrag/Fors-
    varsmaktens-vardegrund-/. Accessed April 25, 2006.
Foucault, Michel. 1979. *The History of Sexuality*. London: Allen Lane.
Fox Keller, Evelyn. 1985. *Reflections on Gender and Science*. New Haven, CT: Yale Univer-
    sity Press.
Fox, Mary-Jane. 2001. "The Idea of Women in Peacekeeping: Lysistrata and Antigone."
    In *Women and International Peacekeeping*, ed. Louise Olsson and Torunn L. Tryg-
    gestad, 9–22. Portland, OR: Cass.
French Presidency. 2008. *Review of the Implementation by the Member States and the EU
    Institutions of the Beijing Platform for Action—Indicators Concerning Women and
    Armed Conflicts*, 16596/08 ADD 1–2. Brussels: Council of the European Union.
Friis, Karsten. 1999. *Stat, nasjon og verneplikt* [State, nation and conscription]. NUPI
    Rapport 246. Oslo: Norsk utenrikspolitisk Institutt.
Gardiner, Judith Kegan, ed. 2001. *Masculinity Studies & Feminist Theory*. New York:
    Colombia University Press.
"Gay Dutch Soldiers Responsible for Srebrenica Massacre Says U.S. General." 2010. *The
    Telegraph*, March 19. http://www.telegraph.co.uk/news/worldnews/northameri-
    ca/usa/7478738/Gay-Dutch-soldiers-responsible-for-Srebrenica-massacre-says-
    US-general.html. Accessed June 5, 2011.
Genderforce. 2007. "Conquering a New Field: Integrating the Gender Dimension into
    International Missions." Stockholm: EQUAL Community Initiative. http://ec.
    europa.eu/employment_social/equal/practical-examples/opport-06-se-gender-
    force_en.cfm. Accessed April 8, 2011.
Göbel, Henrietta. 2000. Interview by author SAF, Stockholm. November 15.
Goetschel, Laurent. 1999. "Neutrality, a Really Dead Concept?" *Cooperation and Conflict*
    34(2): 115–139.
Goldstein, Joshua. 2001. *War and Gender*. Cambridge, UK: Cambridge University Press.
Göteborgsposten. 2008. "Riksarkivet tar snoppstrid." February 21. Göteborg, Sweden:
    Author. http://www.gp.se/nyheter/sverige/1.186735-riksarkivet-tar-snoppstrid. Ac-
    cessed February 21, 2008.
Grady, Kate. 2010. "Sexual Exploitation and Abuse by UN Peacekeepers: A Threat to
    Impartiality." *International Peacekeeping*, 17(2): 215–228.
Griffith, James. 2007. "Further Considerations Concerning the Cohesion-Performance
    Relation in Military Settings." *Armed Forces & Society*, 34(1): 138–147.
"Gripens rykte hänger i luften." 2011. *Svenska Dagbladet*, April 1. http://www.svd.se/
    naringsliv/gripens-rykte-hanger-i-luften_6056827.svd. Accessed April 1, 2011.
Gruber, James. 1998. "The Impact of Male Work Environments and Organizational
    Policies on Women's Experiences of Sexual Harassment." *Gender & Society*, 12(3):
    301–320.
Gustafsson, Bengt. 1995. "Widening Horizons in Swedish Security Analysis?" In *New
    Thinking in International Relations: Swedish Perspectives*, ed. Rutger Lindahl and
    Gunnar Sjöstedt, 121–137. Stockholm: Swedish Institute of International Af-
    fairs.
Gutek, Barbara. 1989. "Sexuality in the Workplace: Key Issues in Social Research and
    Organizational Practice." In *The Sexuality of Organization*, ed. Jeff Hearn, Debo-
    rah L. Sheppard, Peter Tancred, and Gibson Burrell, 56–70. Newbury Park, CA:
    SAGE.
Gya, Giji, Charlotte Isaksson, and Marta Martinelli. 2009. "Report on ESDP Missions
    in the Democratic Republic of the Congo." Paper presented at: From Commit-

ment to Action—The EU Delivering to Women in Conflict and Post-Conflict, conference organized by the French Presidency of the Council of the European Union, the Commission, and UNIFEM. Brussels, October 10, 2008.

Haaland, Torunn Laugen. 2011. "Friendly War-Fighters and Invisible Women: Perceptions of Gender and Masculinities in the Norwegian Armed Forces on Missions Abroad." In *Making Gender, Making War: Violence, Military, and Peacekeeping Practices*, ed. Annica Kronsell and Erika Svedberg, 63–75. New York: Routledge.

Hafner-Burton, Emilie, and Mark Pollack. 2002. "Gender Mainstreaming and Global Governance." *Feminist Legal Studies*, 10: 285–298.

Hall, Rodney Bruce. 1999. *National Collective Identity—Social Constructs and International Systems*. New York: Columbia University Press.

Hansen, Lene. 2001. "Gender, Nation, Rape: Bosnia and the Construction of Security." *International Feminist Journal of Politics*, 3(1): 55–75.

Hansen, Wibeke, Oliver Ramsbotham, and Tom Woodhouse. 2001. *Berghof Handbook for Conflict Transformation*. Berlin: Berghof Research Center for Constructive Conflict Management.

Harding, Sandra. 1991. *Whose Science? Whose Knowledge? Thinking from Women's Lives*. Buckingham, UK: Open University Press.

Hartsock, Nancy C.M. 1983. *Money, Sex and Power—Toward a Feminist Historical Materialism*. Boston: Northeastern University Press.

Hassan, Pahwasha. 2010. Speech. Delivered at the Women, Peace and Security: The Afghan View conference. Tallinn, Estonia, November 11–12.

Hearn, Jeff. 2004. "From Hegemonic Masculinity to the Hegemony of Men." *Feminist Theory*, 5(1): 49–72.

———. 2011. "Sexualities, Work, Organizations, and Managements: Empirical, Policy and Theoretical Challenges." In *Handbook of Gender, Work and Organization*, ed. Emma Jeanes, David Knights, and Patricia Yancey Martin. Chichester, UK: Wiley.

Hearn, Jeff, Deborah Sheppard, Pet Tancred-Sheriff, and Gibson Burrell, eds. 1989. *The Sexuality of Organization*. London: SAGE.

Hearn, Jeff, and Wendy Parkin. 1995. *"Sex" at "Work": The Power and Paradox of Organisation Sexuality*. New York: St. Martin's.

Held, David. 1995. *Democracy and the Global Order: From the Modern State to Cosmopolitan Governance*. Stanford, CA: Stanford University Press.

Higate, Paul. 2004. *Gender and Peacekeeping: Case Studies: The Democratic Republic of the Congo and Sierra Leone*. ISS Monograph No. 91. Paris: Institute for Security Studies.

———. 2007. "Peacekeepers, Masculinities, and Sexual Exploitation," *Men and Masculinities*, 10(1): 99–119.

———. 2002. "Traditional Gendered Identities: National Service and the All-Volunteer Force." In *The Comparative Study of Conscription in the Armed Forces*, Comparative Social Research, Vol. 20, ed. Lars Mjøset and Stephen Van Holde, 229–235. Oxford, UK: Elsevier Science.

Higate, Paul, and John Hopton. 2005. "War, Militarism and Masculinities." In *Handbook of Studies on Men & Masculinities*, ed. Michael Kimmel, Jeff Hearn, and Raewyn Connell, 432–447. Thousand Oaks, CA: SAGE.

Higate, Paul, and Marsha Henry. 2004. "Engendering (In)security in Peace Support Operations." *Security Dialogue*, 35(4): 481–498.

———. 2010. "Space, Performance and Everyday Security in the Peacekeeping Context." *International Peacekeeping*, 17(2): 32–48.

Hirdman, Yvonne. 1989. *Att lägga livet till rätta: studier i svensk folkhemspolitik.* Stockholm: Carlsson.

———. 1998. *Med kluven tunga. LO och genusordningen.* Stockholm: Atlas.

HOF Board. 2010. "Homo-, bi-, and oche transpersoner i Försvaret." http://www.hof.org.se/index.html. Accessed August 30, 2011.

Holmström, Mikael. 2011. *Den Dolda Alliansen. Sveriges hemliga NATO-förbindelser* [The secret alliance. Sweden's secret relations with NATO]. Stockholm: Atlantis.

"Hon riskerar livet varje dag." 2006. *Aftonbladet.* http://www.aftonbladet.se/wendela/article445285.ab. Accessed October 11, 2006.

Hudson, Heidi. 2005. "Peacekeeping Trends and their Gender Implications for Regional Peacekeeping Forces in Africa: Progress and Challenges." In *Gender, Conflict, and Peacekeeping,* ed. Dyan Mazurana, Angela Raven-Roberts, and Jane Parpart, 111–133. Oxford, UK: Rowman & Littlefield.

Huggler, Justin. "Chinese Prostitutes Arrested in Kabul." 2006. *The Independent,* February 10. www.independent.co.uk/news/world/asia/chinese-prostitutes-arrested-in-kabul-restaurant-raids-466118.html. Accessed February 10, 2006.

Hull, Cecilia, Mikael Eriksson, Justin MacDermott, Fanny Rudén, and Annica Waleij. 2009. *Managing Unintended Consequences of Peace Support Operations.* Stockholm: FOI, Swedish Defence Research Agency.

Hutchings, Kimberly. 2008. "Making Sense of Masculinity and War." *Men and Masculinities,* 10(4): 389–404.

Iannello, Kathleen P. 1992. *Decisions Without Hierarchy: Feminist Interventions in Organization Theory and Practice.* New York: Routledge.

Ignatieff, Michael. 2000. *Virtual War: Kosovo and Beyond.* London: Chatto & Windus.

Ingebritsen, Christine. 2002. "Norm Entrepreneurs: Scandinavia's Role in World Politics." *Cooperation and Conflict,* 37(1): 11–23.

———. 2006. *Scandinavia in World Politics.* Lanham, MD: Rowman & Littlefield.

Insats & Försvar. 2006. No. 2. Stockholm: Armed Forces Headquarters.

———. 2006. No. 4. Stockholm: Armed Forces Headquarters.

———. 2006. No. 6. Stockholm: Armed Forces Headquarters.

———. 2007. No. 3. Stockholm: Armed Forces Headquarters.

———. 2007. No. 5. Stockholm: Armed Forces Headquarters.

Ivarsson, Sophia. 2002. *Diskurser kring kvinnor i uniform. Om attityder till kvinnor som officerare och värnpliktiga bland män i försvarsmakten* [Attitudes toward women in uniform]. Stockholm: Försvarshögskolan.

Ivarsson, Sophia, and Anders Berggren. 2001. *Avgångsorsaker bland officerare.* LI Serie T:24. Stockholm: Ledarskapsinstitutionen, Försvarshögskolan.

Ivarsson, Sophia, Anders Berggren, and Nicola Magnusson. 2006. *Könsdiskriminering i den svenska Försvarsmakten. En fördjupad studie avseende förekomsten av sexuella trakasserier och diskriminering pga kön* [Sexual harassment in SAF: An in-depth study on the prevalence of sexual harassment and discrimination due to gender]. Stockholm: Försvarshögskolan.

Ivarsson, Sophia, and Lina Edmark. 2005. "Genusperspektiv på Nordic Battle Group." In *Människan i NBF. Med särskilt focus på internationella insatser,* ed. Anders W. Berggren, 125–153. Stockholm: Försvarshögskolan.

———. 2007. *Utlandsstyrkans internationella insatser ur ett genusperspektiv, hinder och möjligheter för implementering av FN resolution 1325* [International operations from a gender perspective, obstacles and opportunities for the implementation of 1325]. Stockholm: Försvarsskolan.

Jabri, Vivienne. 1996. *Discourses on Violence*. Manchester, UK: Manchester University Press.

———. 2007. *War and the Transformation of Global Politics*. Basingstoke, UK: Palgrave Macmillan.

Jacobsson, Mats. 1998. *Man eller Monster—Kustjägarnas mandomsprov* [Man or monster—Trial of manhood among commando soldiers]. Nora, Sweden: Nya Doxa.

Jacoby, Wade, and Christopher Jones. 2008. "The EU Battle Groups in Sweden and the Czech Republic: What National Defense Reforms Tell Us about European Rapid Reaction Capabilities." *European Security*, 17(2): 315–338.

Jacquot, Sophie. 2010. "The Paradox of Gender Mainstreaming: Unanticipated Effects of New Modes of Governance in the Gender Equality Domain." *West European Politics*, 33(1): 118–135.

Jakobsen, Peter Viggo. 2007. "Still Punching Above Their Weight? Nordic Cooperation in Peace Operations after the Cold War." *International Peacekeeping*, 14(4): 458–475.

Janowitz, Morris. 1983. *The Reconstruction of Patriotism: Education for Civic Consciousness*. Chicago: University of Chicago Press.

Jennings, Kathleen. 2010. "Unintended Consequences of Intimacy: Political Economies of Peacekeeping and Sex Tourism." *International Peacekeeping*, 17(2): 229–243.

Jennings, Kathleen, and Vesna Nikolic-Ristanovic. 2009. *UN Peacekeeping Economies and Local Sex Industries: Connections and Implications*. MICROCON Research Working Paper 17. Brighton, UK: MICROCON.

Jensen, Carsten. 2009. "Nedstigning i skuggriket." *Dagens Nyheter*, January 25.

Joenniemi, Pertti. 2006a. "Introduction: Unpacking Conscription." In *The Changing Face of European Conscription*, ed. Pertti Joenniemi, 1–12. Aldershot, UK: Ashgate.

———, ed. 2006b. *The Changing Face of European Conscription*. Aldershot, UK: Ashgate.

Johansson, Per-Ola, Susanne Lindholm, and Kenth Johansson. 2009. *Marinen behöver Dig! Rapport efter möte med tjänslediga yrkesofficerare som är kvinnor* [The Marine Corps need you! Meeting female officers on leave]. Karlskrona, Sweden: Försvarsmakten, Marinen.

Johnson, Chalmers A. 2004. *The Sorrows of Empire: Militarism, Secrecy and the End of the Republic*. London: Verso.

Jukarainen, Pirjo. 2011. "Men Making Peace in the Name of Just War—The Case of Finland?" In *Making Gender, Making War: Violence, Military, and Peacekeeping Practices*, ed. Annica Kronsell and Erika Svedberg, 90–103. New York: Routledge.

Jufors, Madeleine. 2008. Telephone interview with author. September 1.

Jung Fiala, Irene. 2008. "Unsung Heroes: Women's Contributions in the Military and Why Their Song Goes Unsung." In *Women in the Military and in Armed Conflict*, ed. Helena Carreiras and Gerhard Kümmel, 49–61. Wiesbaden, Germany: VS Verlag für Sozialwissenschaften.

Kaldor, Mary. 2001. *New & Old Wars: Organized Violence in a Global Era*. Stanford, CA: Stanford University Press.

Kammarens protokoll. 2004. *Protokoll av interpellationsdebatten* [Protocol of the Parliamentary debate]. 2004/05:48. Stockholm: Riksdagen [Swedish Parliament].

Kanetake, Machiko. 2010. "Whose Zero Tolerance Counts? Reassessing a Zero Tolerance Policy against Exploitation and Abuse by UN Peacekeepers." *International Peacekeeping*, 17(2): 200–214.

Kaplan, Laura Duhan. 1994. "Women as Caretaker: An Archetype that Supports Patriarchal Militarism." *Hypatia*, 9(2): 123–133.

Katzenstein, Mary Fainsod. 1998. *Faithful and Fearless: Moving Feminist Protest inside the Church and Military*. Princeton, NJ: Princeton University Press.

Kaufman, Joyce P., and Kristen P. Williams. 2007. *Women, the State, and War: A Comparative Perspective on Citizenship and Nationalism*. Lanham, MD: Lexington.

Kerttunen, Mika, Tommi Koivula, and Tommy Jeppson. 2005. "EU Battlegroups. Theory and Development in the Light of Finnish-Swedish Co-operation." Research Report 30. Helsinki: Department of Strategic and Defence Studies.

Kier, Elizabeth. 1998. "Homosexuals in the U.S. Military: Open Integration and Combat Effectiveness." *International Security*, 23(2): 5–39.

Kimmel, Michael, Jeff Hearn, and Raewyn Connell, eds. 2005. *Handbook of Studies on Men & Masculinities*. Thousand Oaks, CA: SAGE.

King, Anthony. 2007. "The Existence of Group Cohesion in the Armed Forces." *Armed Forces & Society*, 33(4): 638–645.

———. 2006. "The Word of Command: Communication and Cohesion in the Military." *Armed Forces & Society*, 32(4): 493–512.

Kirke, Charles. 2009. "Group Cohesion, Culture, and Practice." *Armed Forces & Society*, 35(4): 745–753.

Kite, Cynthia. 2006. "The Domestic Background: Public Opinion and Party Attitudes Towards Integration in the Nordic Countries." In *The Nordic Countries and the European Security and Defence Policy*, ed. Alyson J.K. Bailes, Gunilla Herolf, and Bengt Sundeliua, 99–109. Stockholm: SIPRI.

Knapp, Deborah Erdos. 2008. "Ready or Not? Homosexuality, Unit Cohesion, and Military Readiness." *Employee Responsibilities and Rights Journal*, 20(4): 227–247.

Körlof, Björn. 2001. "Värnplikt, försvarsvilja och försvarsdebatt." [Conscription, the will to defend the nation and the defense debate] Anförande av generaldirektör Björn Körlof vid Kungliga Krigsvetenskapsakademiens symposium: "Värnplikten 100 år." September 4.

Kramer, Zachary. 2010. "Heterosexuality and Military Service." *Northwestern University Law Review Colloquy*, 104: 341–365.

Kronsell, Annica. 2005. "Gendered Practices in Institutions of Hegemonic Masculinity: Reflections from Feminist Standpoint Theory." *International Feminist Journal of Politics*, 7(2): 280–298.

———. 2006. "Methods for Studying Silences: Gender Analysis in Institutions of Hegemonic Masculinity." In *Feminist Methodologies for International Relations*, ed. Brooke Ackerly, Maria Stern, and Jacqui True, 108–128. Cambridge, UK: Cambridge University Press.

———. 2009. "Negotiations in Networks. The Importance of Personal Relations and Homosociality." In *Diplomacy in Theory and Practice*, ed. Karin Aggestam and Magnus Jerneck, 241–256. Malmö, Sweden: Liber.

———. (2010). "Gender and Governance." *The International Studies Encyclopedia*. Denemark, Robert A. 2671–2695 Hoboken, NJ: Wiley-Blackwell.

Kronsell, Annica, and Erika Svedberg. 2007. "Folkförankring på riktigt? Kön, medborgarskap och demokrati i de frivilliga försvarsorganisationerna" [Gender, citizenship and democracy in the voluntary defense organizations]. Unpublished manuscript. Krisberedskapsmyndigheten, Stockholm.

———. 2001. "The Duty to Protect: Gender in the Swedish Practice of Conscription." *Cooperation and Conflict*, 36(2): 153–176.

————. 2006. "The Swedish Military Manpower Policies and their Gender Implications." In *The Changing Face of European Conscription*, ed. Pertti Joenniemi, 137–160. Aldershot, UK: Ashgate.

————, eds. 2011. *Making Gender, Making War: Violence, Military and Peacekeeping Practices*. New York: Routledge.

Kryhl, Tomas. 1996. "Kvinnan tar befälet: vägen till kvinnliga officerare i svenskt försvar" [Women in command: Female officers in the Swedish defense]. *Militärhistorisk tidskrift*, 2: 7–57.

Kümmel, Gerhard. 2008. "Chivalry in the Military." In *Women in the Military and in Armed Conflict*, ed. Helena Carreiras and Gerhard Kümmel, 183–199. Wiesbaden, Germany: VS Verlag für Sozialwissenschaften.

Kvande, Elin. 1999."In the Belly of the Beast: Constructing Femininities in Engineering Organizations." *The European Journal of Women's Studies*, 6(3): 305–328.

Kwon, Insook. 2000. "A Feminist Exploration of Military Conscription: The Gendering of the Connections Between Nationalism, Militarism and Citizenship in South Korea." *International Feminist Journal of Politics*, 3(1): 26–54.

Lagergren, Fredrika. 1999. *På andra sidan välfärdsstaten* [On the other side of the welfare state]. Stockholm: Symposion.

Leander, Anna. 2004. "Drafting Community: Understanding the Fate of Conscription." *Armed Forces & Society*, 30(4): 571–599.

Leander, Anna, and Pertti Joenniemi. 2006. "Conclusion: National Lexica of Conscription." In *The Changing Face of European Conscription*, ed. Pertti Joenniemi, 161–174. Aldershot, UK: Ashgate.

Lees, Sue. 1986. *Loosing Out: Sexuality and Adolescent Girls*. London: Hutchinson.

Levi, Margaret. 1997. *Consent, Dissent and Patriotism*. Cambridge, UK: Cambridge University Press.

Levy, Yagil. 2007. "The Right to Fight: A Conceptual Framework for the Analysis of Recruitment Policy toward Gays and Lesbians." *Armed Forces & Society*, 33(1): 186–202.

Liebert, Ulrike. 2002. "Europeanising Gender Mainstreaming: Constraints and Opportunities in the Multilevel Euro-Polity." *Feminist Legal Studies*, 10: 241–256.

Lindstrom, Gustav. 2007. "Enter the EU Battlegroups." Chaillot Paper 97. Paris: Institute for Security Studies.

Lipman-Blumen, Jean. 1976. "Toward a Homosocial Theory of Sex Roles: An Explanation of the Sex Segregation of Social Institutions." *Signs*, 1(3): 15–31.

Lister, Ruth. 1997. *Citizenship: Feminist Perspectives*. London: Macmillan.

Locher, Birgit. 2007. *Trafficking in Women in the European Union: Norms, Advocacy Networks and Policy Change*. Wiesbaden, Germany: VS Verlag für Sozialwissenschaften.

Locher, Brigit, and Elisabeth Prügl. 2001. "Feminism and Constructivism: Worlds Apart or Sharing the Middle Ground?" *International Studies Quarterly*, 45(1): 111–129.

Logue, John. 1989. "The Legacy of Swedish Neutrality." In *The Committed Neutral: Sweden's Foreign Policy*, ed. Bengt Sundelius, 35–65. Boulder, CO: Westview.

Lutz, Catherine, Matthew Gutman, and Keith Brown. 2009. "Conduct and Discipline in UN Peacekeeping Operations: Culture, Political Economy and Gender." Report submitted to the Conduct and Discipline Unit, UN Department of Peacekeeping Operations. Providence, RI: Brown University, Watson Institute of International Studies.

Lykke, Nina. 2010. *Feminist Studies. A Guide to Intersectional Theory, Methodology and Writing*. New York: Routledge.

MacDoun, Robert, Elizabeth Kier, and Aaron Belkin. 2006. "Does Social Cohesion Determine Motivation in Combat?" *Armed Forces & Society*, 32(4): 646–654.

MacKinnon, Catherine A. 1989. *Toward a Feminist Theory of the State*. Cambridge, MA: Harvard University Press.

Magnusson, Peter. 1998. *Organisationskulturer i Försvarsmakten* [Organizational cultures in the Armed Forces]. LI Serie F:10. Stockholm: Försvarshögskolan.

Maninger, Stephan. 2008. "Women in Combat: Reconsidering the Case against the Deployment of Women in Combat-Support and Combat Units." In *Women in the Military and in Armed Conflict*, ed. Helena Carreiras and Gerhard Kümmel, 9–27. Wiesbaden, Germany: VS Verlag für Sozialwissenschaften.

Mann, Michael. 1987. "The Roots and Contradictions of Modern Militarism." *New Left Review*, 162: 35–50.

Manners, Ian. 2002. "Normative Power Europe: A Contradiction in Terms?" *Journal of Common Market Studies*, 40(2): 235–258.

———. 2006. "Normative Power Europe Reconsidered: Beyond the Crossroads." *Journal of European Public Policy*, 13(2): 182–199.

Mänskliga rättigheter i Afghanistan [Human rights in Afghanistan]. 2007. Stockholm: Utrikesdepartementet, Regeringskansliet.

March, James G., and Johan P. Olsen. 1989. *Rediscovering Institutions: The Organizational Basis of Politics*. New York: Free Press.

Mared, Sarah. 2008. *Det som inte syns. Om homosexualitet i de militära källorna från 1900 talets början* [That which is silent. Homosexuality in military sources since the beginning of the 20th century]. Stockholm: Riksarkivet.

"Margot Wallström Fed Up with EU 'Reign of Old Men.'" 2008. EUobserver.com. February 8. http://euobserver.com/9/25631?rss_rk=1.

Mazey, Sonia. 2002. "Gender Mainstreaming Strategies in the EU: Delivering on an Agenda?" *Feminist Legal Studies* 10: 227–240.

Mazurana, Dyan, Angela Raven-Roberts, and Jane Parpart, eds. 2005. *Gender, Conflict, and Peacekeeping*. Oxford, UK: Rowman & Littlefield.

McCarry, Melanie. 2007. "Masculinity Studies and Male Violence: Critique or Collusion?" *Women's Studies International Forum*, 30: 404–415.

McEnaney, Laura. 2000. *Civil Defense Begins at Home: Militarization Meets Everyday Life in the Fifties*. Princeton, NJ: Princeton University Press.

Mehler, Andreas. 2008. "Positive, Ambiguous or Negative? Peacekeeping in the Local Security Fabric." In *Critical Currents*, 5: 41–64. Occasional Paper Series. Uppsala: Dag Hammarskjöld Foundation.

Meola, Lynn. 1997. "Sexual Harassment in the Army." In *Wives & Warriors: Women and the Military in the U.S. and Canada*, ed. Laurie Weinstein and Christie White, 145–149. Westport, CT: Bergin & Garvey.

Mersiades, Michael. 2005. "Peacekeeping and Legitimacy: Lessons from Cambodia and Somalia." *International Peacekeeping*, 12(2): 205–221.

Miles, Lee. 2006. "Domestic Influences on Nordic Security and Defence Policy: From the Perspective of Fusion." In *The Nordic Countries and the European Security and Defence Policy*, ed. Alyson J.K. Bailes, Gunilla Herolf, and Bengt Sundelius, 77–98. Stockholm: SIPRI.

Mjøset, Lars, and Stephen Van Holde. 2002. "Killing for the State, Dying for the Nation: An Introductory Essay on the Life Cycle of Conscription into Europe's Armed Forces." In *The Comparative Study of Conscription in the Armed Forces*, Comparative Social Research, Vol. 20, ed. Lars Mjøset and Stephen Van Holde, 3–94. Oxford, UK: Elsevier Science.

Moberg, Eva. 2008. "Vad stort sker, sker tyst." *Fred och Frihet, IKFF* [Journal of Women's International League for Peace and Freedom], 1: 5–6.

Moelker, René, and Jolanda Bosch. 2008. "Women in the Netherlands Armed Forces." In *Women in the Military and in Armed Conflict*, ed. Helena Carreiras and Gerhard Kümmel, 81–127. Wiesbaden, Germany: VS Verlag für Sozialwissenschaften.

Møller, Bjørn. 2002. "Conscription and its Alternatives." In *The Comparative Study of Conscription in the Armed Forces*, Comparative Social Research, Vol. 20, ed. Lars Mjøset and Stephen Van Holde, 277–306. Oxford, UK: Elsevier Science.

Moon, Katharine. 1997. *Sex Among Allies. Military Prostitution in U.S –Korea Relations*, New York: Colombia University Press.

Mosse, George L. 1990. *Fallen Soldiers: Reshaping the Memory of the World Wars*. New York: Oxford University Press.

Mukwege, Denis. 2009. "Rape as a War Strategy." Paper presented at the Second Swedish-Dutch Conference on Gender Equality: Women in War Zones. The Hague, December 3.

Münkler, Herfried. 2005. *The New Wars*. Cambridge, UK: Polity.

Munn, Jamie. 2008. "The Hegemonic Male and Kosovar Nationalism, 2000–2005." *Men and Masculinities*, 10(4): 440–456.

Myrttinen, Henri. 2003. "Disarming Masculinities." *Disarmament Forum* 4: 37–46. http://www.unidir.org/pdf/articles/pdf-art1996.pdf. Accessed April 24, 2011.

Nasuti, Matthew. "Illicit Sex and Boozing by Civilian U.S. Contractors and Officials in Afghanistan Undermining Mission." 2009. Kabul Press, November 15. http://kabulpress.org/my/spip.php?article4262. Accessed November 15, 2009.

Nationella sekretariatet för Genusforskning. 2009. "Jämställdhetsintegrering i statliga mydigheters verksamhet" [Gender mainstreaming governmental agencies]. Rapport 2/09. Göteborg, Sweden: Göteborgs universitet.

Naughton, Philippe. "Dutch Outrage as U.S. General Blames Gay Soldiers for Srebrenica." 2010. *The Sunday Times*, March 19. http://www.timesonline.co.uk/tol/news/world/us_and_americas/article7068523.ece. Accessed June 5, 2011

Nilsson, Anne. 1990. "Kvinnan som officer" [The woman as officer]. *Försvar i Nutid*, 90: 20. Stockholm: Centralförbundet Folk och Försvar.

Nye, Joseph S. 2004. *Soft Power. The Means to Success in World Politics*. New York: Public Affairs.

Olsson, Louise, and Johan Tejpar. 2009. "Operational Effectiveness and UN Resolution 1325: Practices and Lessons from Afghanistan." User Report. Stockholm: Swedish Defence Research Agency (FOI).

Olsson, Louise, and Martin Åhlin. 2009. "Strengthening ESDP Missions and Operations through Training on UNSCR 1325 and 1820." Report from EU seminar organized by the Swedish Presidency, Brussels, July. http://www.consilium.europa.eu/homepage.aspx. Accessed April 1, 2011.

Olsson, Louise, and Torunn L. Tryggestad, eds. 2001. *Women and International Peacekeeping*. Portland, OR: Cass.

Olsson, Sven E. 1990. *Social Policy and Welfare State in Sweden*. Lund, Sweden: Arkiv.

Osburn, C. Dixon, and Michelle M. Benecke. 1997. "Conduct Unbecoming: Second Annual Report on 'Don't Ask, Don't Tell, Don't Pursue.'" In *Wives & Warriors: Women and the Military in the U.S. and Canada*, ed. Laurie Weinstein and Christie White, 151–177. Westport, CT: Bergin & Garvey.

Otto, Dianne. 2007. "Making Sense of Zero Tolerance Policies in Peacekeeping Sexual Economies." *Sexuality and the Law: Feminist Engagements*, ed. Vanessa Munro and Carl Stychin, 259–282. New York: Routledge-Cavendish.

Ottosson, Jenny. 1997. "Rapport: Undersökning av värnpliktiga kvinnors situation i Försvarsmakten 1996/97" [Female conscripts in SAF]. Projekt Tjej-2000. Stockholm: Försvarsmakten.

"Ove Bring: Jas-beslutet inte kontroversiellt." 2011. *Dagens Nyheter*, March 29. http://www.dn.se/nyheter/sverige/ove-bring-beslutet-inte-kontroversiellt. Accessed March 29, 2011.

Parliamentary Memo. 2004. *Homosexuella i Försvaret—läget i några europeiska länder*. Dnr. 2004:1288. Stockholm: Swedish Parliament.

Paris, Roland. 2002. "International Peacebuilding and the *Mission Civilisatrice*." *Review of International Studies*, 28: 637–656.

Pateman, Carole. 1988. *The Sexual Contract*. Stanford, CA: Stanford University Press.

Patomäki, Heikki. 2000. "Beyond Nordic Nostalgia: Envisaging a Social-Democratic System of Global Governance." *Cooperation and Conflict*, 35(2): 115–154.

Pattison, James. 2008. "Humanitarian Intervention and a Cosmopolitan UN Force." *Journal of International Political Theory*, 4(1): 126–145.

Penttinen, Elina. 2011. "Nordic Women and International Crisis Management: A Politics of Hope?" In *Making Gender, Making War: Violence, Military, and Peacekeeping Practices*, ed. Annica Kronsell and Erika Svedberg, 153–165. New York: Routledge.

"Percentages of Female Soldiers in NATO Countries` Armed Forces." 2007. Washington, DC: Office on Women in the NATO Forces and Women's Research and Education Institute. http://www.nato.int/issues/women_nato/perc_fem_soldiers.jpg. Accessed June 10, 2009.

Perspektivstudien 2009. "Rapport från persspektivstudien. Det militärstrategiska utfallsrummet" [Sweden's military strategic context]. Unpublished manuscript. Swedish Armed Forces, Stockholm.

Peters, Guy B. 2005. *Institutional Theory in Political Science. The "New" Institutionalism*. 2d ed. London: Continuum.

Peterson V. Spike. 2008. "'New Wars' and Gendered Economies." *Feminist Review*, 88: 7–20.

Peterson, V. Spike, and Jacqui True. 1998. "New Times and New Conversations." In *The "Man" Question in International Relations*, ed. Marysia Zalewski and Jane Parpart, 14–27. Boulder, CO: Westview.

Pettersson, Lena. 2008. "Positiv särbehandling av underrepresenterat kön vid antagningen till officersprogrammet 2007" [Affirmative action for the officer program 2007]. Unpublished manuscript. The Defense College, Stockholm

Pettersson, Lena, and Alma Persson. 2005. *"Tål man inte jargongen kan man väl inte kriga": Kvinna och yrkesofficer i den svenska Försvarsmakten* [Experiencing being a female professional officer in SAF]. Stockholm: Arbetslivsinstitutet.

Phillips, Anne. 2000. *Närvarons politik-den politiska representationen av kön, etnicitet och ras*. Lund, Sweden: Studentlitteratur.

Pouligny, Béatrice. 2006. *Peace Operations Seen from Below. UN Missions and Local People*. London: Hurst.

Proposition 2006/07:1. 2006. "Budgetpropositionen för 2007" [Budget bill for 2007]. Stockholm: The Government's office.

Proposition 1993/94:147. 1993. "Delad makt—delat ansvar" [Shared power—shared responsibility]. Stockholm: Ministry of Health and Social Affairs.

Proposition 2001/02:10. 2001. "Fortsatt förnyelse av Totalförsvaret" [Rejuvenating the defense]. Stockholm: Ministry of Health and Social Affairs.

Proposition 2004/05:5. 2004."Vårt framtida försvar" [Our future defense]. Stockholm: Ministry of Health and Social Affairs.

Proposition 2008/09:140. 2008."Ett användbart försvar" [A useful defense]. Stockholm: Ministry of Health and Social Affairs.

"Prostitution under the Rule of Taliban." 1999. The Revolutionary Association of Women of Afghanistan (RAWA). http://www.rawa.org/rospi.htm. Accessed June 10, 2010.

"Prövas inte i strid nu heller." 2011. Sydsvenskan, April 8. http://www.sydsvenskan.se/sverige/article1436000/Provas-inte-i-strid-nu-heller.html. Accessed April 8, 2011.

Prügl, Elisabeth. 2009. "Does Gender Mainstreaming Work?" International Feminist Journal of Politics, 11(2): 174–195.

Puechguirbal, Nadine. 2010. "Discourses on Gender, Patriarchy and Resolution 1325: A Textual Analysis of UN Documents." International Peacekeeping, 17(2): 172–187.

Pyke, D. Karen. 1996. "Class-Based Masculinities. The Interdependence of Gender, Class and Interpersonal Power." Gender & Society, 10(5): 527–549.

Qadiry, Tahir. 2008. "Under Wraps, Prostitution Rife in North Afghanistan." Reuters, May 20. http://www.alertnet.org/thenews/newsdesk/ISL173658.htm. Accessed June 5, 2010.

Reardon, Betty. 1996. Sexism and the War System. Syracuse, NY: Syracuse University Press.

Rees, Teresa. 2005. "Reflections on the Uneven Development of Gender Mainstreaming in Europe." International Feminist Journal of Politics, 7(4): 555–574.

"Repeal of 'Don't Ask, Don't Tell' Paves Way for Gay Sex Right on Battlefield, Opponents Fantasize." 2010. The Onion, July 12. http://www.theonion.com/articles/repeal-of-dont-ask-dont-tell-paves-way-for-gay-sex,17698/. Accessed July 12, 2010.

Rhen, Elisabeth, and Ellen Johnson Sirleaf. 2002. Women, War and Peace: The Independent Experts' Assessment on the Impact of Armed Conflict on Women and Women's Role in Peace-Building. New York: United Nations Development Fund for Women.

Riemer, Jeffrey. 1998. "Durkheim's 'Heroic Suicide' in Military Combat." Armed Forces & Society, 25(1): 103–120.

Riley, Robin, Chandra Talpade Mohanty, and Minnie Bruce Pratt. 2008. Feminism and War: Confronting U.S. Imperialism. New York: Zed Books.

Risse, Tomas. 2004. "Global Governance and Communicative Action." Government and Opposition, 39(2): 288–313.

Ritter, Carl. 2008. "World Takes Notice of Swedish Prostitute Laws." The Independent, March 18. www.independent.co.uk/news/world/europe/world-takes-notice-of-swedish-prostitute-laws-796793.html. Accessed March 19, 2008.

Ritter, Gretchen. 2002. "Of War and Virtue: Gender, American Citizenship and Veterans' Benefits after WW II." In The Comparative Study of Conscription in the Armed Forces, Comparative Social Research, Vol. 20, ed. Lars Mjøset and Stephen Van Holde, 201–228. Oxford, UK: Elsevier Science.

Röhr, Ulrike, Meike Spitzner, Elisabeth Stiefel, and Uta Winterfeld. 2008. "Gender Justice as the Basis for Sustainable Climate Policies: A Feminist Background Paper." Berlin: German NGO Forum on Environment and Development.

Rubinstein, Robert A., Diana M. Keller, and Michael E. Scherger. 2008. "Culture and Interoperability in Integrated Missions." *International Peacekeeping*, 15(4): 540–555.

Ruddick, Sarah. 1989. *Maternal Thinking: Toward a Politics of Peace*. Boston: Beacon.

Ryan, Alan. 2004. "Cosmopolitan Objectives and the Strategic Challenges of Multinational Military Operations." In *Forces for Good? Cosmopolitan Militaries in the Twenty-First Century*, ed. Lorraine Elliott and Graeme Cheeseman, 65–78. Manchester, UK: Manchester University Press.

Sainsbury, Diane, and Christina Bergqvist. 2009. "The Promise and Pitfalls of Gender Mainstreaming: The Swedish Case." *International Feminist Journal of Politics*, 11(2): 216–234.

Sasson-Levy, Orna. 2003. "Feminism and Military Gender Practices: Israeli Women Soldiers in 'Masculine' Roles." *Sociological Inquiry*, 73(3): 440–465.

Scales, Ann. 1989. "Militarism, Male Dominance and Law: Feminist Jurisprudence as Oxymoron?" *Harvard Women's Law Journal*, 12: 25–73.

"Security Council Demands Immediate and Complete Halt to Acts of Sexual Violence against Civilians in Conflict Zones, Unanimously Adopting Resolution 1820." 2008. New York: United Nations Security Council. www.un.org/News/Press/docs/2008/sc9364.doc.htm. Accessed November 26, 2009.

Seifert, Ruth. 1996. "The Second Front: The Logic of Sexual Violence in Wars." *Women's Studies International Forum*, 19(1–2): 35–43.

Shaw, Martin. 2005. *The New Western Way of War*. Cambridge, UK: Polity.

Shepherd, Laura. 2008. *Gender, Violence & Security*. New York: Zed Books.

Shiels, Edward, and Morris Janowitz. 1948. "Cohesion and Disintegration in the Wehrmacht in World War II." *Public Opinion Quarterly*, 12: 279–315.

Siebold, Guy. 2007. "The Essence of Military Group Cohesion." *Armed Forces & Society*, 33(2): 286–295.

Simic, Olivera. 2010. "Does the Presence of Women Really Matter? Towards Combating Male Sexual Violence in Peacekeeping Operations." *International Peacekeeping*, 17(2): 188–199.

Simonsen, Sven Gunnar. 2007. "Building 'National' Armies—Building Nations? Determinants of Success for Postintervention Integration Efforts." *Armed Forces & Society*, 33(4): 571–590.

Sinclair, Amanda. 2005. *Doing Leadership Differently: Gender, Power and Sexuality in a Changing Business Culture*. 2d ed. Victoria, Australia: Melbourne University Press.

Sion, Liora. 2008. "Peacekeeping and the Gender Regime. Dutch Female Peacekeepers in Bosnia and Kosovo." *Journal of Contemporary Ethnography*, 37(5): 561–585.

———. 2006. "'Too Sweet and Innocent for War?' Dutch Peacekeepers and the Use of Violence." *Armed Forces & Society*, 32(3): 454–474.

SIPRI Arms Transfers Database. 2009. Stockholm: Stockholm International Peace Research Institute. http://www.sipri.org/databases/armstransfers. Accessed April 27, 2010.

Sjoberg, Laura. 2006. *Gender, Justice, and the Wars in Iraq. A Feminist Reformulation of Just War Theory*. Lanham, MD: Lexington.

Skjelsbaek, Inger. 2007. "Gender Aspects of International Military Interventions: National and International Perspectives." PRIO Papers. Oslo: International Peace Research Institute.

———. 2001. "Is Femininity Inherently Peaceful? The Construction of Femininity in the War." In *Gender, Peace & Conflict*, ed. Inger Skjelsbaek and Dan Smith, 47–67. Thousand Oaks, CA: SAGE.

Skjelsbaek, Kjell. 1979. "Militarism: Its Dimensions and Corollaries: An Attempt at Conceptual Clarification." *Journal of Peace Research*, 16(3): 213–229.

Skrivelse 2007/08:51. 2008. "Nationell strategi för svenskt deltagande i internationella freds och säkerhetsbefrämjande verksamhet" [National strategy for participation in international peace and security missions]. Stockholm: Swedish Parliament. http://www.riksdagen.se/Webbnav/index.aspx?nid=37&rm=2007/08&bet=51&typ=prop. Accessed April 21, 2011.

Smith, Dan. 2001. "The Problem of Essentialism." In *Gender, Peace & Conflict*, ed. Inger Skjelsbaek and Dan Smith, 32–46. Thousand Oaks, CA: SAGE.

Smith, William. 2007. "Anticipating a Cosmopolitan Future: The Case of Humanitarian Military Intervention." *International Politics*, 44: 72–89.

Sørensen, Henning. 2000. "Conscription in Scandinavia During the Last Quarter Century: Developments and Arguments." *Armed Forces & Society*, 26(2): 313–334.

"Sorg i lägret efter svenskens död." 2010. *Svenska Dagbladet*, October 17.

SOU (Official Report of the Swedish Government). 1990:89. *En ny värnpliktslag* [A new conscription law]. Stockholm: Försvarsdepartementet.

———. 1977:26. *Kvinnan och Försvarets Yrken* [Women and the military profession]. Stockholm: Försvarsdeparementet, Fritzes förlag.

———. 1994:11. *Om kriget kommit. Förberedelser för mottagandet av militärt bistånd 1949–1969* [In case of war. Preparations for receiving military support]. Stockholm: Försvarsdepartementet.

———. 2001:23. *Personal för ett nytt försvar* [Staffing the new defense]. Stockholm: Försvarsdepartementet.

———. 1992:139. *Totalförsvarsplikt* [Total defense duty]. Stockholm: Försvarsdepartementet.

———. 2009:63. *Totalförsvarsplikt och frivillighet* [Total defense duty and voluntarism]. Stockholm: Försvarsdepartmentet. http://www.sweden.gov.se/sb/d/11664/a/128859.

———. 2000:21. *Totalförsvarsplikten* [The total defense duty]. Stockholm: Försvarsdepartementet.

———. 1965:68. *Värnplikten* [Conscription]. Stockholm: Försvarsdepartementet.

———. 1984:71 *Värnplikten i Framtiden*, [Conscription in the Future] Försvarsdepartementet, Stockholm.

Speech Sälen. 2006. Introductory speech by the Defense Minister Leni Björklund at the Annual National Conference hosted by the Organization Society and Defence, in Sälen, Sweden.

Stamnes, Eli. 2007. "Introduction." *International Peacekeeping*, 14(4): 449–457.

"Statement of Government Policy" 2002. October 1. Stockholm: Ministry of Foreign Affairs. http://www.regeringen.se/sb/d/241/a/30413. Accessed October 1, 2002.

"Statement of Government Policy" 2006. February 15. Stockholm: Ministry of Foreign Affairs. http://www.regeringen.se/sb/d/10229/a/18985. Accessed August 25, 2011.

"Statement of Government Policy in the Parliamentary Debate on Foreign Affairs." 2010. February 17. Stockholm: Ministry of Foreign Affairs and Government Offices of Sweden. http://www.sweden.gov.se/sb/d/5304/a/139731. Accessed June 6, 2010.

"Statement of Government Policy in the Parliamentary Debate on Foreign Affairs." 2011. February 16. Stockholm: Ministry of Foreign Affairs and Government Offices of Sweden. http://www.regeringen.se/sb/d/5298/a/139694, Accessed August 19, 2011.

Steans, Jill. 1997. *Gender and International Relations: An Introduction*. Oxford, UK: Polity.

Stiehm, Judith Hicks. 1989. *Arms and the Enlisted Woman*. Philadelphia: Temple University Press.

———. 1982. "The Protected, the Protector, the Defender." *Women's Studies International Forum* 5(3/4): 367–376.

Stiglmayer, Alexandra, ed. 1994. *Mass Rape: The War against Women in Bosnia-Herzegovina*. Lincoln: University of Nebraska Press.

Stratigaki, Maria. 2005. "Gender Mainstreaming vs. Positive Action: An Ongoing Conflict in EU Gender Equality Policy." *European Journal of Women's Studies* 12(2): 165–186.

Stütz, Göran. 2006. *Opinion 2006: Den svenska allmänhetens syn på samhället, säkerhetspolitiken och försvaret* [Swedish public opinion on security and defense policy]. Stockholm: Styrelsen för Psykologiskt Försvar.

———. 2008. *Opinion 2008: Den svenska allmänhetens syn på samhället, säkerhetspolitiken och försvaret* [Swedish public opinion on security and defense policy]. Stockholm: Styrelsen för Psykologiskt Försvar.

Sundelius, Bengt. 1989a."Committing Neutrality in an Antagonistic World." In *The Committed Neutral: Sweden's Foreign Policy*, ed. Bengt Sundelius, 1–13. Boulder, CO: Westview.

———, ed. 1989b. *The Committed Neutral: Sweden's Foreign Policy*, Boulder, CO: Westview.

Sundevall, Fia. 2011. *Det sista manliga yrkesmonopolet. Genus och militärt arbete i Sverige 1865–1989* [The last male bastion in the labour market: Gender and military work in Sweden 1865–1989]. Göteborg, Sweden: Makadam.

"Svenska soldater beskjutna i Afghanistan." 2006. *Dagens Nyheter*, September 10.

"Svenska soldater i eldstrid i Afghanistan." 2006. *Dagens Nyheter*, September 12.

"Svenskar Under Attack." 2006. *Dagens Nyheter*, July 1.

Swedish Armed Forces (SAF). 2005. *Årsredovisning* [Annual report]. Stockholm: Author.

———. 2006a. *Årsredovisning* [Annual report]. Stockholm: Author.

———. 2007. *Årsredovisning* [Annual report]. Stockholm: Author.

———. 2009. *Årsredovisning* [Annual report]. Stockholm: Author.

———. 1996. *Den kreativa olikheten. Information om steg 2* [The creative difference, step 2]. SAF internal document.

———. 1995. *Den kreativa olikheten. Redovisning av bakgrund och arbetsläge* [The creative difference: Background]. SAF internal document.

———. 2006b. *Försvarsmaktens Jämställdhetsplan 2006–2008* [SAF equality plan 2006–2008]. Stockholm: Author.

———. 2009. *Försvarsmaktens Jämställdhetsplan 2009–2011* [SAF equality plan 2009–2011]. Stockholm: Author.

———. 2009. Information leaflet on gender and operational effectiveness.

Swedish Presidency. 2009. "Beijing +15: The Platform for Action and the European Union." Report from the Swedish Presidency of the Council of the European Union. se2009.eu. ec.europa.eu/social/BlobServlet?docId=4336&langId=en. Accessed August 31, 2011.

Sylvester, Christine. 1994. "Empathetic Cooperation: A Feminist Method for IR." *Millennium: Journal of International Studies*, 23(2): 315–334.

Syrén, Håkan. 2004. *Vägen framåt—en liten bok om en stor förändring* [The way forward—A small book about a great change]. Stockholm: Försvarsmakten.

Thomas, Alison M. 1997. "Men Behaving Badly? A Psychosocial Exploration of the Cultural Context of Sexual Harassment." In *Sexual Harassment—Contemporary Feminist Perspectives*, ed. Alison M. Thomas and Celia Kitzinger, 131–153. Buckingham, UK: Open University Press.

Tickner, Ann J. 2006. "Feminism Meets International Relations: Some Methodological Issues." In *Feminist Methodologies for International Relations*, ed. Brooke Ackerly, Maria Stern, and Jacqui True, 19–41. Cambridge, UK: Cambridge University Press.

Tilly, Charles. 1990. *Coercion, Capital and European States, AD 990–1990*. Oxford, UK: Blackwell.

Tolgfors, Sven. 2009. Radio interview, SR P1 Swedish Radio, September 19.

Tornbjer, Charlotte. 2002. "Den nationella modern. Moderskap i konstruktioner av svensk nationell gemenskap under 1900-talets första hälft" [Motherhood in the construction of the Swedish national identity]. Ph.D. diss., Historiska institutionen vid Lunds universitet, Lund, Sweden.

Trägårdh, Lars. 1997. "Statist Individualism: On the Culturality of the Nordic Welfare State." In *The Cultural Construction of Norden*, ed. Øystein Sørensen and Bo Stråth, 253–285 Oslo: Scandinavian University Press.

Tronto, Joan. 2008. "Is Peacekeeping Care Work? A Feminist Reflection on 'The Responsibility to Protect.'" In *Global Feminist Ethics: Feminist Ethics and Social Theory*, ed. Rebecca Whisnant and Peggy DesAutels, 179–199. Lanham, MD: Rowman & Littlefield.

United Nations General Assembly. 2005. "Investigation by the Office of Internal Oversight Services into Allegations of Sexual Exploitation and Abuse in the UN Mission in the Democratic Republic of Congo." Report (A/59/661) January. New York: Author.

United Nations Peacekeeping Department. 2008a. "United Nations Mission in Ethiopia and Eritrea." New York: Author. http://www.un.org/en/peacekeeping/missions/past/unmee/. Accessed April 27, 2011.

———. 2008b. "United Nations Peacekeeping Operations." New York: Author. http://www.un.org/en/peacekeeping/. Accessed April 27, 2011.

United Nations Security Council. 2008. 5916th meeting, Thursday, 19 June 2008, 10 a.m. New York. http://www.un.org/Depts/dhl/resguide/scact2008.htm. Accessed August 27, 2011.

United Nations Secretariat. 2003. "Secretary-General's Bulletin: Special Measures for Protection from Sexual Exploitation and Sexual Abuse." October 9. ST/SGB/2003/13. New York: Author.

U.S. Department of State. 2009. "2008 Human Rights Report: Afghanistan." Washington, DC: Author. http://www.state.gov/g/drl/rls/hrrpt/2008/sca/119131.htm. Accessed November 28, 2009.

"U.S. Judge Orders an End to 'Don't Ask, Don't Tell' Military Policy." 2010. *The Guardian*, October 12. http://www.guardian.co.uk/world/2010/oct/12/us-judge-dont-ask-dont-tell-gay-army. Accessed October 12, 2010.

Valenius, Johanna. 2007a. "A Few Kind Women: Gender Essentialism and Nordic Peacekeeping Operations." *International Peacekeeping*, 14(4): 510–523.

———. 2007b. "Gender Mainstreaming in ESDP Missions." Chaillot Paper 101. Paris: Institute for Security Studies.

Van Creveld, Martin. 2001. *Men, Women and War*. London: Cassell.

———. 2008. *The Culture of War*. New York: Ballantine Books.

Van der Vleuten, Anna. 2007. *The Price of Gender Equality. Member States and Govern-*
ance in the European Union. Aldershot, UK: Ashgate.
"Varför är vi i Afghanistan?" 2009. *Dagens Nyheter,* February 5.
Värnpliktsnytt [The conscript's journal]. 1989. (December). Stockholm: Department of
Defence.
———. 1991. (October). Stockholm: Department of Defence.
———. 1995. (November). Stockholm: Department of Defence.
———. 1996. (November). Stockholm: Department of Defence.
———. 1998. (February). Stockholm: Department of Defence.
Värnpliktsrådet. 2008a. "Förbandsrapport F17" [Report from visit at F17 military
unit]. November 14. http://www.varnpliktsradet.se/index.php?page=arkiv&cho
ice=view&aid=2.Accessed July 20, 2009.
———. 2008b. "Rapport från generalbesök på AMF1 [Report from a visit at the am-
phibian battalion]. 20080829. http://www.varnpliktsradet.se/index.php?page=
arkiv&choice=view&aid=2.Accessed July 20, 2009.
———. 2009. "Välkommen till verkligheten" [Welcome to reality]. Rapport från värnpliktsrådet. http://www.varnpliktsradet.se/index.php?page=arkiv&choice=view
&aid=2. Accessed July 20, 2009.
Väyrynen, Tarja. 2004. "Gender and UN Peace Operations: The Confines of Modernity."
International Peacekeeping, 11(1): 125–142.
Verloo, Mieke. 2001. "Another Velvet Revolution? Gender Mainstreaming and the Politics of Implementation." IWM Working Paper 5/2001. Vienna: Institute for Human Sciences.
Viktorin, Mattias. 2005. "The New Military: From National Defence and Warfighting
to International Intervention and Peacekeeping." *Statsvetenskaplig Tidskrift,*
107(3): 259–277.
Von Sydow, Björn. 2001. Interview at Grand Hotel, Lund by author. November 27.
Wadley, Jonathan. 2010. "Gendering the State: Performativity and Protection in Inter-
national Security." In *Gender and International Security: Feminist Perspectives,* ed.
Laura Sjoberg, 38–58. New York: Routledge.
Wæver, Ole. 1999. "Identity, Communities and Foreign Policy: Discourse Analysis as
Foreign Policy Theory." In *Between Nations and Europe: Regionalism, Nationalism
and the Politics of Union,* ed. Lene Hansen and Ole Wæver, 20–49. London:
Routledge.
Walby, Sylvia. 2005. "Gender Mainstreaming: Productive Tensions in Theory and Prac-
tice." *Social Politics: International Studies in Gender, State & Society,* 12(3): 321–343.
———. 2009. *Globalization & Equalities. Complexity and Contested Modernities.* Thousand Oaks, CA: SAGE.
Walzer, Michael. 1977. *Just and Unjust Wars. A Moral Argument with Historical Illustra-*
tions. New York: Basic Books.
Wedin, Lars. 2006. "The Impact of EU Capability Targets and Operational Demands on
Defence Concepts and Planning: The Case of Sweden." In *The Nordic Countries and
the European Security and Defence Policy,* ed. Alyson J.K. Bailes, Gunilla Herolf,
and Bengt Sundeliua, 141–149. Stockholm: SIPRI.
Weibull, Louise. 2001. *Tjejmönstring—Lyckad rekrytering eller lockad rekryt? En studie av
en ny rekryteringsdrive inom försvarsmakten* [A study of a new recruiting process
in SAF]. LI Serie F:18. Stockholm: Försvarshögskolan.
Weinstein, Laurie, and Christie C. White, eds. 1997. *Wives & Warriors: Women and the
Military in the U.S. and Canada.* Westport, CT: Bergin & Garvey.

Willett, Susan. 2010. "Introduction: Security Council Resolution 1325: Assessing the Impact on Women, Peace and Security." *International Peacekeeping*, 17(2): 142–158.

Whitehead, Stephen. 2002. *Men and Masculinities: Key Themes and New Directions*. Cambridge, UK: Polity.

Whitworth, Sandra. 2004. *Men, Militarism & UN Peacekeeping. A Gendered Analysis*. Boulder, CO: Lynne Rienner.

———. 2008. "Militarized Masculinity and Post-Traumatic Stress Disorder." In *Rethinking the Man Question. Sex, Gender and Violence in International Relations*, ed. Jane L. Parpart and Marysia Zalewski, 109–126. New York: Zed Books.

Wollinger, Susanne. 2000. *Mannen i ledet. Takt och otakt i värnpliktens skugga* [The conscripted man]. Stockholm: Carlsson.

Woodward, Alison. 2004. "European Gender Mainstreaming: Promises and Pitfalls of Transformative Policy." In *Equity in the Workplace: Gendering Workplace Policy Analysis*, ed. Heidi Gottfried and Laura Ann Reese, 77–100 Lanham, MD: Lexington.

———. 2008. "Too Late for Gender Mainstreaming? Taking Stock in Brussels." *Journal of European Social Policy*, 18(3): 289–302.

Woodward, Rachel, and Patricia Winter. 2004. "Discourses of Gender in the Contemporary British Army." *Armed Forces & Society*, 30(2): 279–301.

Yllner, Nadja. 2010. "Uppdrag granskning. FN arbetarna som köper sex" [Swedish documentary on the Swedish staff on UN missions and their sexual relations with locals]. Sveriges Television, January 27.

Young, Iris Marion. 2007. *Global Challenges: War, Self-Determination and Responsibility for Justice*. Cambridge, MA: Polity.

———. 2000. *Inclusion and Democracy*. Oxford, UK: Oxford University Press.

Yuval-Davis, Nira. 1997. *Gender & Nation*. London: SAGE.

Zalewski, Marysia. 2010. "'I Don't Even Know What Gender Is': A Discussion of the Connections Between Gender, Gender Mainstreaming and Feminist Theory." *Review of International Studies*, 36: 3–27.

Zarkov, Dubravka. 2002. "Srebrenica Trauma: Masculinity, Military and National Self-Image in Dutch Daily Newspapers." In *The Postwar Moment: Militaries, Masculinities and International Peacekeeping*, ed. Cynthia Cockburn and Dubravka Zarkov, 183–203. London: Lawrence & Wishart.

———. 2007. *The Body of War*. Durham, NC: Duke University Press.

Zimmerman, Caroline. "Afghan Sex Trade Thrives Despite Taboos." 2008. *Newser*, June 15. www.newser.com/story/30057/afghan-sex-trade-thrives-despite-taboos.html. Accessed November 28, 2009.

INDEX